D1630182

GOOD HOUSEKEEPING

ENCYCLOPEDIA OF MICROWAVE TECHNIQUES

GOOD HOUSEKEEPING

ENCYCLOPEDIA OF MICROWAVE TECHNIQUES

EBURY PRESS LONDON

Published by Ebury Press
an imprint of The Random Century Group
Random Century House
20 Vauxhall Bridge Road
London SW1V 2SA

Copyright © 1992 The Random Century Group Ltd and
The National Magazine Company Ltd

All rights reserved. No part of this publication may be reproduced, stored in
a retrieval system, or transmitted in any form or by any means, electronic,
mechanical, photocopying, recording, or otherwise, without the prior permission
of the copyright owners.

ISBN 0 09175389 9

The expression GOOD HOUSEKEEPING as used in the title
of this book is the trade mark of The National
Magazine Company Limited and The Hearst Corporation, registered
in the United Kingdom and the USA, and other
principal countries in the world, and is the absolute
property of The National Magazine Company Limited and
The Hearst Corporation. The use of this trade mark
other than with the express permission of The National
Magazine Company Limited or The Hearst Corporation is strictly
prohibited.

Editor: Felicity Jackson
Designer: Graham Dudley
Photographers: Simon Butcher, Paul Kemp, David
Johnson, James Murphy, Alan Newnham

NOTE: WE DO NOT RECOMMEND THE USE OF RECYCLED PAPER PRODUCTS
IN MICROWAVE COOKING SINCE THEY MAY CONTAIN IMPURITIES WHICH
COULD CAUSE SPARKS.

Filmset by Advanced Filmsetters (Glasgow) Ltd
Printed and bound in Italy by New Interlitho S.p.a., Milan

CONTENTS

MICROWAVE COOKERS

GENERAL RECIPE NOTES

Follow either metric or imperial measures for the recipes in this book; they are not interchangeable.

Bowl sizes
Small bowl = about 900 ml (1½ pints)
Medium bowl = about 2.3 litres (4 pints)
Large bowl = about 3.4 litres (6 pints)

Covering
Cook, uncovered, unless otherwise stated.
At the time of going to press, it has been recommended by the Ministry of Agriculture, Fisheries and Food that the use of cling film should be avoided in microwave cooking. When a recipe requires you to cover the container, either cover with a lid or a plate.

Size 2 eggs should be used unless otherwise stated.

REHEATING

Foods can be reheated in a microwave with no loss of colour or flavour. Furthermore, an entire course for a meal can be reheated on the same plate that it is served on. When doing this, keep the height of the various items as even as possible and arrange the more dense and thicker items towards the outside of the plate.

Follow the same procedure as when reheating foods conventionally and cover foods such as vegetables to prevent them drying out. Stir the food occasionally for even heating; items that cannot be stirred should be returned or re-arranged.

Reheating in a microwave is extremely quick so special attention should be given to small items of food to avoid overcooking. Special attention must be paid to cooked pastry and breads. Place these on absorbent kitchen paper to absorb moisture during reheating and prevent the bottom from becoming soggy. Microwaves are attacted to the moist fillings in pies and pasties, so that the liquid will heat up quickly. The steam produced by this is often absorbed into the pastry covering which may not leave it as crisp as that reheated in a conventional oven.

Only general guidelines concerning the time it takes to reheat food can be given, as so much depends on the type of food and the initial temperature of the food you are reheating. As a general rule, for an individual serving, try starting with 2 minutes on HIGH, test to see if the food is hot and then repeat in 2-minute bursts if it is not. With pastry foods such as fruit pies, the outer pastry should feel just warm. The temperature of pastry and filling will equalise if given a few minutes standing time.

COMBINED OVEN OWNERS

Combined ovens combine conventional and microwave methods of cooking so that food browns as well as cooking quickly. One of the disadvantages of cooking in a microwave cooker is that baked dishes do not brown or crisp. In this book, we show you how to overcome these disadvantages, but if you own a combined cooker you will not have these problems. In this case, you should follow your manufacturers' instructions.

How to use the Recipes in this Book with your Cooker Settings

Unlike conventional ovens, the power output and heat controls on various microwave cookers do not follow a standard formula. When manufacturers refer to a 700-watt cooker, they are referring to the cooker's POWER OUTPUT; its INPUT, which is indicated on the back of the cooker, is double that figure. The higher the wattage of a cooker, the faster the rate of cooking, thus food cooked at 700 watts on full power cooks in half the time of food cooked at 350 watts. That said, the actual cooking performance of one 700-watt cooker may vary slightly from another with the same wattage because factors such as cooker cavity size affect cooking performance. The vast majority of microwave cookers sold today are either 600, 650 or 700 watts, but there are many cookers still in use which may be 400 and 500 watts.

IN THIS BOOK
HIGH refers to 100%/full power output
of 600–700 watts.
MEDIUM refers to 60% of full power.
LOW is 35% of full power.

Whatever the wattage of your cooker, the HIGH/FULL setting will always be 100% of the cooker's output. Thus your highest setting will correspond to HIGH.

However, the MEDIUM and LOW settings used in this book may not be equivalent to the MEDIUM and LOW settings marked on your cooker. As these settings vary according to power input, use the following calculation to estimate the correct setting for a 600–700 watt cooker. This simple calculation should be done before you use the recipes for the first time, to ensure successful results. Multiply the percentage power required by the total number of settings on your cooker and divide by 100. To work out what setting MEDIUM and LOW correspond to on your cooker, use the following calculation.

$$Medium\ (60\%)$$
= %Power required
× Total Number
of Cooker Settings
$$\div 100 = Correct\ Setting$$
$$= \frac{60 \times 9}{100} = 5$$

$$Low\ (35\%)$$
= %Power required
× Total Number
of Cooker Settings
$$\div 100 = Correct\ Setting$$
$$= \frac{35 \times 9}{100} = 3$$

If your cooker power output is lower than 600 watts, then you must allow a longer cooking and thawing time for all recipes and charts in this book.

Add approximately 10–15 seconds per minute for a 500 watt cooker, and 15–20 seconds per minute for a 400 watt cooker. No matter what the wattage of your cooker is, you should always check food before the end of cooking time, to ensure that it does not get overcooked. Don't forget to allow for standing time.

7

CHAPTER 1

BOILING

As in conventional cooking, boiling is the method of cooking food in boiling liquid. The foods that are usually cooked in this way are vegetables, pasta and rice. However, because microwave cooking is a naturally moist form of cooking, less liquid is usually needed than in conventional cooking. Never attempt to boil eggs in the microwave as they will explode.

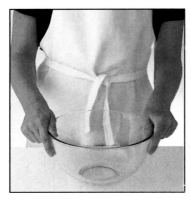

1 Use a large container. This allows liquids to boil without spilling over the top and allows space for stirring.

2 Never fill any bowl more than two-thirds full.

3 If you require more than 300 ml ($\frac{1}{2}$ pint) water, then it is quicker and more economical to cover the food with boiling water.

4 Cover the container with a large plate. Cling film is no longer recommended for use in a microwave.

5 Cover the rice, pasta and pulses by about 2.5 cm (1 inch) of boiling water then cover and cook for the recommended time.

6 Cut vegetables into uniformly sized pieces so that they cook evenly when boing boiled.

CREAMY PORRIDGE
———————SERVES 2———————
▲

A basic porridge to flavour as you like—try clear honey, demerara sugar, chopped dried dates or prunes, or top with a spoonful of Greek yogurt and toasted nuts. If you prefer salty porridge simply add salt to taste with the oats and the milk.

50 g (2 oz) porridge oats   300 ml ($\frac{1}{2}$ pint) creamy milk

1 PUT THE OATS and the milk in a medium heatproof serving bowl and cook on HIGH for 4–5 minutes until boiling and thickened, stirring frequently.

2 STIR IN THE flavouring of your choice (see introduction) and serve immediately.

GREEN SPLIT PEA SOUP
———————SERVES 4–6———————
▲

Large quantities of water are slow to heat up in the microwave, therefore it is quicker to boil the water in a kettle and then pour it over the split peas.

175 g (6 oz) dried green split peas   15 ml (1 tbsp) olive oil
2 leeks, finely chopped and washed   freshly grated nutmeg
2 celery sticks, trimmed and finely chopped salt and pepper
2 medium carrots, finely chopped   chopped fresh parsley, to garnish
1 garlic clove, skinned and crushed

1 PUT THE PEAS in a large bowl and pour over enough water to cover. Leave to soak overnight.

2 THE NEXT DAY put the leeks, celery, carrots, garlic and oil in a large bowl. Cover and cook on HIGH for 5 minutes.

3 DRAIN THE PEAS and add to the vegetables. Pour in 900 ml (1½ pints) boiling water and mix well together. Cover and cook on HIGH for 25 minutes until the peas are very soft, stirring occasionally.

4 WHEN THE SPLIT peas are cooked, turn the mixture into a blender or food processor and process until smooth. Season with nutmeg and salt and pepper to taste and pour into an ovenproof serving bowl.

5 RETURN THE SOUP to the cooker and cook on HIGH for 3 minutes or until the soup is hot. Garnish with chopped parsley and serve the soup immediately with warm wholemeal rolls.

VICHYSSOISE
SERVES 4
▲

*Though made from the humble potato, nothing could make a more
sophisticated starter to a summer meal than this chilled leek and potato
soup. A good, rich home-made stock is ideal but if you have to fall back
on a stock cube to save time, use less of the cube than recommended
and do not add salt.*

50 g (2 oz) butter or margarine	1 litre (1¾ pints) hot chicken stock
4 leeks, trimmed and sliced	salt and pepper
1 medium onion, skinned and sliced	200 ml (7 fl oz) single cream
2 potatoes, peeled and sliced	fresh chives, to garnish

1 PUT THE BUTTER or margarine in a large bowl and cook on HIGH for
1 minute until melted. Add the leeks and onion, cover and cook on HIGH for
5–7 minutes until softened.

2 ADD THE POTATOES, stock, salt and pepper to taste and cook on HIGH for
15–17 minutes until the vegetables are very soft, stirring frequently.

3 ALLOW TO COOL slightly, then rub through a sieve or purée in a blender
or food processor until smooth. Pour into a large serving bowl and stir in
the cream. Chill for at least 4 hours.

4 To SERVE, WHISK the soup to ensure an even consistency. Pour into
individual bowls and snip fresh chives on top to garnish.

MIXED VEGETABLE SOUP
SERVES 6
▲

*Make this simple soup when you have lots of vegetables to use up.
Use any vegetables at hand but try to include a good mixture for
maximum flavour, such as onions, leeks, carrots, parsnips, potatoes,
swede, celery, turnip or celeriac—the possibilities are endless. Chop
them all finely, to reduce the cooking time. The easiest way to do this is
in a food processor. Add a final enrichment of cream or yogurt.*

1.1 kg (2½ lb) mixed vegetables (see above)	salt and pepper
2 bay leaves	150 ml (¼ pint) double cream, crème fraiche,
bouquet garni	soured cream or single cream (optional)

1 FINELY CHOP THE vegetables and put in a large bowl with the bay leaves,
bouquet garni and 1.4 litres (2½ pints) water.

2 COVER AND COOK ON HIGH for 30–40 minutes or until the vegetables are
very soft. Rub through a sieve or purée in a blender or food processor until
smooth.

3 SEASON TO TASTE with salt and pepper. Return the soup to a clean ovenproof serving bowl and reheat on HIGH for 3–4 minutes or until hot. Stir in cream, if desired.

CHILLED COURGETTE, MINT AND YOGURT SOUP
SERVES 4–6

▲

A light, low-calorie soup, perfect for the weight conscious. Choose tender, young courgettes for the best flavour and do not be tempted to peel them as most of the flavour is in the skin.

1 medium onion, skinned and finely chopped
1 large potato, peeled and grated
600 ml (1 pint) vegetable or chicken stock
450 g (1 lb) courgettes, trimmed and coarsely grated

30 ml (2 tbsp) chopped fresh mint
150 ml ($\frac{1}{4}$ pint) natural yogurt
salt and pepper
courgette slices and sprigs of mint, to garnish

1 PUT THE ONION, potato and half of the stock in a large bowl. Cover and cook on HIGH for 8–10 minutes or until very soft.

2 ADD THE COURGETTES and continue to cook on HIGH for 4 minutes or until the courgettes are soft.

3 RUB THROUGH A sieve or purée in a blender or food processor until smooth. Add the remaining stock, the mint and the yogurt and season to taste with salt and pepper.

4 COVER AND CHILL in the refrigerator for at least 4 hours before serving, garnished with courgette slices and mint sprigs.

SPINACH SOUP
————SERVES 4————
▲

When buying fresh spinach choose bright, green leaves.

15 ml (1 tbsp) vegetable oil
1 large onion, skinned and chopped
450 g (1 lb) fresh spinach, washed, trimmed
  and roughly chopped or 225 g (8 oz)
  frozen leaf spinach
15 ml (1 level tbsp) flour

600 ml (1 pint) boiling chicken or vegetable
  stock
freshly grated nutmeg
salt and pepper
60 ml (4 tbsp) natural yogurt

1 PUT THE OIL and onion in a medium bowl. Cover and cook on HIGH for 5–7 minutes until softened.

2 ADD THE SPINACH, re-cover and cook on HIGH for 3–4 minutes, or 8–9 minutes until thawed if using frozen spinach, stirring occasionally.

3 SPRINKLE IN THE flour and cook on HIGH for 30 seconds, then gradually stir in the stock. Season with nutmeg and salt and pepper to taste. Cook on HIGH for about 4 minutes until boiling, stirring occasionally.

4 LEAVE TO COOL slightly, then purée the soup in a blender or food processor. Pour the soup back into the bowl and cook on HIGH for 2 minutes until boiling. Ladle the soup into warmed bowls and swirl a spoonful of yogurt into each before serving.

WALNUT SOUP
————SERVES 4–6————
▲

Although this soup may sound unusual, it is quite delicious!

1 garlic clove
175 g (6 oz) walnuts
600 ml (1 pint) chicken stock

150 ml ($\frac{1}{4}$ pint) single cream
salt and pepper

1 SKIN THE GARLIC and put into a food processor or blender with the walnuts. Work until finely crushed. If using a blender you may need to add a little stock. Very gradually pour in the stock and blend until smooth.

2 POUR THE SOUP into a large bowl and cook on HIGH for 8–10 minutes until boiling, stirring occasionally.

3 STIR IN THE cream, reserving about 60 ml (4 tbsp) and season to taste with salt and pepper. Serve hot, with a swirl of cream on top.

Chilled Courgette, Mint and Yogurt Soup (page 11)

BEAN AND OLIVE PÂTÉ
SERVES 4–6 as a starter

▲

Certain beans can be cooked more quickly in the microwave cooker if cooking up to 225 g (8 oz), but they still require overnight soaking. Serve this well flavoured pâté with strips of pitta bread warmed in the cooker on HIGH for 1–1½ minutes.

100 g (4 oz) black-eye beans
175 g (6 oz) black olives, stoned
1 garlic clove, skinned and crushed
30 ml (2 tbsp) natural yogurt
15 ml (1 tbsp) lemon juice

15 ml (1 tbsp) olive oil
large pinch of ground cumin
salt and pepper
black olives, to garnish

1 PUT THE BEANS in a large bowl and pour over enough water to cover. Leave to soak overnight.

2 THE NEXT DAY, drain the beans, return to the bowl and pour over enough boiling water to cover by about 2.5 cm (1 inch). Cover and cook on HIGH for 30–40 minutes or until tender.

3 DRAIN THE BEANS and transfer to a food processor or blender. Add the remaining ingredients, seasoning to taste with salt and pepper, and purée until smooth.

4 TURN INTO A serving bowl and leave to cool. Serve garnished with a few olives.

PASTA WITH SPINACH AND RICOTTA
SERVES 2 as a main course

▲

Both spinach and ricotta cheese are used extensively in Italian cooking and are often cooked with pasta. Low-fat ricotta is made from the whey left over from cheese production.

450 g (1 lb) fresh spinach, washed or 225 g
   (8 oz) frozen leaf spinach
225 g (8 oz) fresh green tagliatelle
salt and pepper
30 ml (2 tbsp) olive oil
1 garlic clove, skinned and crushed

1 small onion, skinned and finely chopped
100 g (4 oz) ricotta cheese
5 ml (1 tsp) chopped fresh marjoram
   (optional)
freshly grated nutmeg

1 IF USING FRESH spinach, shred the leaves into small pieces. Set aside.

2 PUT THE TAGLIATELLE and salt to taste in a large bowl and pour over enough boiling water to cover by about 2.5 cm (1 inch). Cover and cook on HIGH for 3–4 minutes until just tender. Drain and toss with half of the oil.

14

3 PUT THE REMAINING oil, garlic and onion in a medium bowl. Cover and cook on HIGH for 4–5 minutes or until softened, stirring occasionally.

4 STIR IN THE spinach and cook on HIGH for 5 minutes if using fresh spinach and 7–8 minutes if using frozen, stirring occasionally. Set aside.

5 CRUMBLE THE RICOTTA cheese into the pasta. Add the spinach mixture, marjoram if using and nutmeg and pepper to taste. Toss together, then cook on HIGH for 2–3 minutes to heat through. Serve immediately with a green salad.

TAGLIATELLE WITH MUSHROOMS AND TWO CHEESES

————————————SERVES 2 as a main course————————————

▲

Stilton cheese is marvellous for cooking, and gives a very distinctive flavour to the mushroom sauce in this pasta dish. Mozzarella, used here for the topping, is a famous Italian cheese. It is deliciously creamy when melted.

225 g (8 oz) fresh tagliatelle
salt and pepper
25 g (1 oz) butter or margarine
1 garlic clove, skinned and crushed
225 g (8 oz) mushrooms, thinly sliced

50 g (2 oz) Stilton cheese
60 ml (4 tbsp) double cream
1 egg, lightly beaten
100 g (4 oz) Mozzarella cheese

1 PUT THE TAGLIATELLE and salt to taste in a large bowl and pour over enough boiling water to cover by about 2.5 cm (1 inch). Cover and cook on HIGH for 3–4 minutes until just tender. Leave to stand, covered. Do not drain.

2 MEANWHILE, PUT THE butter or margarine, garlic and mushrooms in a large bowl, cover and cook on HIGH for 3–4 minutes, until the mushrooms are softened, stirring occasionally. Crumble in the Stilton cheese, then stir in the cream and cook on HIGH for 2 minutes, stirring once.

3 DRAIN THE PASTA and season with lots of pepper. Mix into the mushroom sauce. Stir in the egg and mix together thoroughly.

4 TURN THE MIXTURE into a buttered flameproof dish and grate the Mozzarella on top. Cook on HIGH for 3–4 minutes or until heated through. Brown the top under a hot grill. Serve with a green salad.

FRESH PASTA WITH COURGETTES AND SMOKED TROUT

————————SERVES 4 as a starter or 2 as a main course————————

▲

*Fresh pasta cooks very quickly and the time will depend on the size
and shape. The timings in this recipe are for tagliatelle; if using shapes
such as shells increase the time by 30 seconds–1 minute.*

2 medium courgettes
15 ml (1 tbsp) olive oil
pinch of saffron
225 g (8 oz) fresh spinach pasta, such as
   tagliatelle
salt and pepper

1 smoked trout, weighing about 225 g (8 oz)
150 ml ($\frac{1}{4}$ pint) crème fraîche or double
   cream
30 ml (2 tbsp) black lumpfish roe
fresh herb sprigs, to garnish

1 CUT THE COURGETTES into very thin diagonal slices. Cut each slice in half.
Put the courgettes, oil and saffron in a medium bowl and cook on HIGH for
1 minute, stirring once.

2 PUT THE SPINACH pasta and salt to taste in a large bowl. Pour over
enough boiling water to cover by about 2.5 cm (1 inch). Cover and cook on
HIGH for 3–4 minutes or until almost tender. Leave to stand, covered, while
finishing the sauce. Do not drain.

3 To FINISH THE sauce, remove and discard the skin and bones from the
trout. Flake the flesh and stir into the courgettes with the crème fraîche or
cream and salt and pepper to taste. Cook on HIGH for 2 minutes until hot
and slightly thickened.

4 DRAIN THE PASTA and return to the large bowl. Pour over the sauce and
toss together to mix. If necessary, reheat the sauce and pasta together on
HIGH for about 2 minutes. Transfer the pasta to four plates, top each with a
spoonful of lumpfish roe and garnish with a herb sprig.

Fresh Pasta with Courgettes and Smoked Trout

SUMMER PASTA
——————SERVES 4 as a light meal——————
▲

*Requiring minimum effort, this is the perfect dish to make on hot,
lazy days in the summer when imported marmande or beefsteak
tomatoes are juicy and red, and fresh basil is plentiful. It is not
essential to do step 1 in advance but it intensifies the flavours and fills
the kitchen with a wonderful basil aroma.*

350 g (12 oz) Brie	salt and pepper
3 large ripe marmande tomatoes	450 g (1 lb) spaghetti
1 large handful of fresh basil leaves	45 ml (3 tbsp) olive oil
2–3 large garlic cloves, skinned and crushed	

1 REMOVE AND DISCARD the thick outer rind from the Brie leaving the top
and bottom rind on. Cut the cheese into small pieces. Coarsely chop the
tomatoes and basil. Carefully mix everything together with the garlic and
season generously with black pepper and a little salt. Cover and leave for
30 minutes–1 hour to let the flavours develop.

2 PUT THE SPAGHETTI and salt to taste in a large bowl and pour over
enough boiling water to cover the pasta by about 2.5 cm (1 inch). Cover
and cook on HIGH for 7–10 minutes or until tender.

3 DRAIN THE PASTA and return to the rinsed out bowl or an ovenproof
serving dish with the oil. Cook on HIGH for 2 minutes or until hot. Add the
cheese and tomato mixture and toss together. Serve immediately.

PASTA SHELLS WITH ANCHOVY
AND PARSLEY DRESSING
————————————SERVES 2 as a light meal————————————
▲

*This is a simple pasta dish using ingredients you're likely to have in
the storecupboard. If you don't have any fresh parsley, omit it rather
than using dried.*

225 g (8 oz) medium dried pasta shells	50 g (2 oz) can anchovy fillets, drained and
salt and pepper	roughly chopped
30 ml (2 tbsp) lemon juice	45 ml (3 tbsp) chopped fresh parsley
1 garlic clove, skinned and crushed	50 g (2 oz) butter

1 PUT THE PASTA and salt to taste in a large bowl. Pour over enough
boiling water to cover by about 2.5 cm (1 inch). Stir, cover and cook on
HIGH for 8 minutes or until almost tender, stirring occasionally. Leave to
stand, covered, while making the dressing. Do not drain.

2 To MAKE THE dressing, put the lemon juice and garlic in a small bowl and stir in the anchovy fillets, parsley and pepper.

3 DRAIN THE PASTA and turn into a warmed serving dish. Toss with the butter. Pour the dressing over and toss together, making sure that all the pasta is coated in dressing. Serve immediately.

FRESH PASTA WITH MUSHROOMS
————————————SERVES 4 as a main course————————————
▲

Mushrooms cook very quickly in the microwave cooker because they contain a lot of water. Try to include a mixture of different varieties in this sauce to give the best flavour.

50 g (2 oz) butter or margarine
1 garlic clove, skinned and crushed
450 g (1 lb) button, cup, flat or oyster
 mushrooms or a mixture
15 ml (1 tbsp) dry vermouth

150 ml ($\frac{1}{4}$ pint) soured cream
salt and pepper
450 g (1 lb) fresh pasta
chopped fresh herbs, to garnish

1 PUT THE BUTTER or margarine and garlic in a large bowl and cook on HIGH for 2 minutes, stirring once.

2 ROUGHLY CHOP ANY large mushrooms and mix into the butter with the vermouth. Cover and cook on HIGH for 3–4 minutes or until the mushrooms are just cooked, stirring once. Stir in the cream and season to taste with salt and pepper. Set aside while cooking the pasta.

3 PUT THE PASTA and salt to taste in a large bowl and pour over enough boiling water to cover by about 2.5 cm (1 inch). Cover and cook on HIGH for 3–4 minutes or until tender.

4 DRAIN THE PASTA and mix into the mushroom mixture. Toss lightly together, then cook on HIGH for 1–2 minutes or until hot. Serve immediately, garnished with fresh herbs.

NOODLES WITH GOAT'S CHEESE AND CHIVES

SERVES 2 as a light meal

▲

*If you don't like goat's cheese, follow the recipe but substitute the
cheese with Feta cheese, cream cheese or grated Cheddar cheese.*

225 g (8 oz) dried egg noodles
salt and pepper
25 g (1 oz) butter or margarine
60 ml (4 tbsp) double cream

75 g (3 oz) fresh goat's cheese, crumbled
30 ml (2 tbsp) snipped fresh chives
snipped fresh chives, to garnish

1 PUT THE NOODLES and salt to taste in a medium bowl. Pour over 900 ml
(1½ pints) boiling water. Stir, cover and cook on HIGH for 3–4 minutes or
until almost tender. Leave to stand, covered. Do not drain.

2 MEANWHILE, PUT THE butter and margarine and cream in a medium
bowl and cook on HIGH for 2 minutes, or until very hot. Stir in the goat's
cheese and season with salt and pepper. Cook on HIGH for 2 minutes.

3 DRAIN THE NOODLES and stir into the cheese mixture with the chives.
Carefully mix together with two forks, then cook on HIGH for 2 minutes or
until hot. Garnish with chives and serve immediately.

RISOTTO ALLA MILANESE

SERVES 4 as a main course

*A true Italian risotto should have a delicious, creamy consistency
while all of the grains of rice are separate and cooked until just al dente.*

75 g (3 oz) butter or margarine
1 small onion, skinned and finely chopped
450 g (1 lb) arborio rice
150 ml (¼ pint) dry white wine
750 ml (1¼ pints) boiling vegetable or
    chicken stock

2.5 ml (½ level tsp) saffron powder or large
    pinch saffron strands
75 g (3 oz) Parmesan cheese, freshly grated
salt and pepper

1 PUT HALF OF the butter or margarine and the onion in a large bowl.
Cover and cook on HIGH for 3–4 minutes or until the onion is softened. Add
the rice, wine, stock and saffron, re-cover and cook on HIGH for 13–15
minutes or until the rice is tender and the water absorbed.

2 STIR IN THE remaining butter or margarine and half of the cheese then
season generously with pepper and a little salt. Serve immediately, with the
remaining Parmesan handed separately.

Summer Strawberry Sorbet (page 29)

BURGUL PILAU
—SERVES 2 as an accompaniment—
▲

Burgul wheat is also known as bulgar or bulghur wheat. There is no difference between them, apart from the spelling.

15 g ($\frac{1}{2}$ oz) flaked almonds
15 g ($\frac{1}{2}$ oz) butter or margarine
1 medium onion, skinned and finely
    chopped
1 garlic clove, skinned and crushed
100 g (4 oz) burgul wheat

200 ml (7 fl oz) boiling chicken stock
salt and pepper
30 ml (2 tbsp) natural yogurt
15 ml (1 tbsp) chopped fresh chives or
    parsley

1 PUT THE ALMONDS and butter or margarine in a medium bowl and cook on HIGH for 1–2 minutes or until the almonds are golden brown, stirring frequently.

2 STIR IN THE onion and garlic, cover and cook on HIGH for 4–5 minutes until softened, stirring occasionally.

3 MEANWHILE, WASH THE burgul wheat in several changes of water. Drain and then stir into the cooked onion and garlic. Stir in the stock, re-cover and cook on HIGH for 6–8 minutes or until tender and all the liquid is absorbed, stirring occasionally.

4 SEASON WELL WITH salt and pepper, then stir in the yogurt and chives or parsley. Serve hot or cold.

HERBY RICE PILAFF
————SERVES 4–6 as an accompaniment————
▲

When serving this well-flavoured brown rice pilaff with a main course dish, choose herbs that complement the main dish.

1 medium onion, skinned and finely
    chopped
1 garlic clove, skinned and crushed
450 ml ($\frac{3}{4}$ pint) boiling vegetable or chicken
    stock

225 g (8 oz) long-grain brown rice
salt and pepper
60 ml (4 tbsp) chopped fresh herbs, such as
    parsley, tarragon, marjoram, chives,
    chervil

1 PUT THE ONION, garlic, stock and rice into a large bowl. Cover, leaving a gap to let steam escape, and cook on HIGH for 30–35 minutes or until tender, stirring once and adding a little extra water if necessary. Leave to stand, covered, for 5 minutes, by which time all the water should be absorbed.

2 SEASON TO TASTE with salt and pepper and mix in the herbs lightly with a fork. Serve immediately.

KEDGEREE

SERVES 4 as a main course

▲

Rice can be cooked by the absorption method (all the added water is absorbed by the rice during cooking) or by the same method as pasta, adding boiling water to cover and then draining after cooking.

225 g (8 oz) long-grain white rice
salt and pepper
450 g (1 lb) smoked haddock fillets
45 ml (3 tbsp) milk
2 eggs, hard-boiled and chopped

50 g (2 oz) butter or margarine
2.5 ml ($\frac{1}{2}$ level tsp) mild curry powder
  (optional)
45 ml (3 tbsp) single cream
45 ml (3 tbsp) chopped fresh parsley

1 PUT THE RICE and salt to taste in a large bowl and pour over enough boiling water to cover by about 2.5 cm (1 inch). Cover and cook on HIGH for 10–12 minutes or until tender. Drain and transfer to an ovenproof dish.

2 PUT THE HADDOCK and the milk in a large shallow dish, cover and cook on HIGH for 4–5 minutes until the fish flakes easily. Flake the fish, discarding the skin. Add the fish to the rice with the cooking liquid and the remaining ingredients. Season to taste with pepper.

3 COOK ON HIGH for 3–4 minutes, stirring occasionally. Serve immediately.

MIXED GRAINS

–SERVES 4–6 as an accompaniment–

▲

Serve this as an accompaniment to make an interesting change from rice or pasta or add grated cheese or vegetables to make a main meal.

50 g (2 oz) butter or margarine
1 medium onion, skinned and chopped
100 g (4 oz) pearl barley
1.1 litres (2 pints) boiling vegetable or
  chicken stock

100 g (4 oz) millet
100 g (4 oz) roasted buckwheat
30 ml (2 tbsp) soy sauce
black pepper

1 PUT THE BUTTER or margarine and the onion in a large bowl. Cover and cook on HIGH for 5–7 minutes or until softened, stirring occasionally.

2 ADD THE BARLEY and half of the stock. Re-cover and cook on HIGH for 20 minutes.

3 ADD THE MILLET, buckwheat and the remaining stock, re-cover and cook on HIGH for 20–25 minutes or until the grains are tender and the water is absorbed.

4 STIR IN THE soy sauce and season to taste with pepper. Leave to stand for 5 minutes to let the flavours mingle, then serve.

CELERIAC AND POTATO PURÉE
————————SERVES 2 as an accompaniment————————
▲

Celeriac has a delicious, delicate celery flavour and aroma which is quite the opposite of its ugly, knobbly appearance. Puréed with potato it makes an interesting accompaniment for fish.

225 g (8 oz) celeriac, peeled
1 medium potato, weighing about 175 g
   (6 oz), peeled
25 g (1 oz) butter or margarine

15 ml (1 tbsp) natural yogurt or milk
salt and pepper
chopped fresh parsley, to garnish

1 CUT THE CELERIAC and potato into 1.5 cm ($\frac{3}{4}$ inch) cubes. Put into a large shallow dish with 60 ml (4 tbsp) water, cover and cook on HIGH for 8–10 minutes until tender, stirring occasionally.

2 DRAIN AND PUT into a blender or food processor with the butter or margarine, yogurt or milk and salt and pepper to taste and purée until just smooth. Return to the dish or turn into an ovenproof serving dish and cook on HIGH for 1–2 minutes until hot. Garnish with parsley and serve immediately.

MINTED CARROTS AND BRUSSELS SPROUTS
————————SERVES 4 as an accompaniment————————
▲

Most vegetables tend to have a crisp texture when cooked in a microwave. If very soft vegetables are required, use the conventional method of cooking.

450 g (1 lb) Brussels sprouts, trimmed
225 g (8 oz) carrots, sliced
50 g (2 oz) butter or margarine

30 ml (2 tbsp) chopped fresh mint
salt and pepper

1 PUT THE SPROUTS and carrots in a large ovenproof serving dish and add 90 ml (6 tbsp) water. Cover and cook on HIGH for 9–12 minutes or until just tender, stirring once during cooking.

2 DRAIN THE VEGETABLES and return to the dish.

3 PUT THE BUTTER or margarine and mint in a small bowl and cook on HIGH for 1 minute or until melted and foaming. Pour the butter over the vegetables and toss until well coated. Season to taste with salt and pepper. Serve hot.

Apricot Cheesecake (page 30)

WARM NEW POTATO SALAD WITH WHOLEGRAIN MUSTARD AND CREAM DRESSING

————————SERVES 2 as an accompaniment————————

▲

New potato salads are definitely best served warm. Don't be tempted to peel or scrub them, the skins add lots of flavour as well as fibre. If you're worried about calories, use natural or Greek yogurt or soured cream instead of the cream.

350 g (12 oz) small new potatoes
10 ml (2 level tsp) wholegrain mustard
75 ml (3 fl oz) double cream

15 ml (1 tbsp) mayonnaise
5 ml (1 tsp) lemon juice
salt and pepper

1 CUT THE POTATOES in half and put in a medium bowl with 60 ml (4 tbsp) water.

2 COVER AND COOK on HIGH for 5–6 minutes until just tender, stirring occasionally.

3 MEANWHILE, MIX THE mustard with the cream, mayonnaise, lemon juice, salt and pepper.

4 DRAIN THE POTATOES and stir into the dressing. Mix together to coat all the potatoes in the dressing. Serve while still warm.

BROAD BEANS WITH BACON

————————SERVES 4–6 as an accompaniment————————

▲

Small, young broad beans are most suited to this method of cooking because they are the most tender. As they get older the beans develop a thin, parchment-like skin which slows down the cooking. If however you cannot find any fresh beans use frozen and cook on HIGH for the same time.

6 streaky bacon rashers
1.1 kg (2½ lb) small young broad beans

25 g (1 oz) butter or margarine
salt and pepper

1 ARRANGE THE BACON on a plate, cover with a sheet of absorbent kitchen paper and cook on HIGH for 5–6 minutes. Quickly remove the paper to prevent it sticking.

2 SHELL THE BEANS and put in a large bowl with 60 ml (4 tbsp) water. Cover and cook on HIGH for 6–10 minutes or until the beans are just tender, stirring occasionally. (The time will depend on the age of the beans.)

3 MEANWHILE, CHOP THE bacon and cut the butter or margarine into small pieces.

4 DRAIN THE BEANS. Add the bacon and butter and toss together so that the butter melts. Season to taste with salt and pepper. Cook on HIGH for 1–2 minutes to reheat if necessary, and serve immediately.

SUMMER STRAWBERRY SORBET
SERVES 2

▲

This sorbet recipe can be adapted for other summer fruits such as raspberries, blackberries or blackcurrants. If using the latter, add a little extra sugar to counteract the sharp fruit.

40 g (1½ oz) sugar
225 g (8 oz) strawberries, halved
finely grated rind and juice of ½ orange

1 egg white, size 6
strawberries or crisp biscuits, to decorate

1 PUT THE SUGAR and 60 ml (4 tbsp) water into a heatproof jug and cook on HIGH for 3–4 minutes or until boiling. Stir until the sugar has dissolved, then cook on HIGH for 4–5 minutes or until reduced to a syrup. Do not stir.

2 MEANWHILE, PUSH THE strawberries through a fine sieve using the back of a wooden spoon. Stir in the orange rind and juice.

3 ALLOW THE SYRUP to cool slightly, then stir into the strawberry purée. Pour into a shallow freezer container, cover and freeze for 1–1½ hours or until just mushy.

4 REMOVE FROM THE freezer, turn into a bowl and beat well with a fork to break down the ice crystals.

5 WHISK THE EGG white until stiff and fold into the sorbet. Pour back into the freezer container and freeze for about 2 hours until firm.

6 To SERVE, COOK ON HIGH for 15–20 seconds to soften slightly. Serve decorated with strawberries or crisp biscuits.

APRICOT CHEESECAKE
————————SERVES 8————————
▲

*Most of the sweetness of this cheesecake comes from the natural
sugar in apricots and orange juice. Other dried fruits such as pears,
dates or peaches work just as well but the colour will not be as good.*

350 g (12 oz) no-soak dried apricots
300 ml (½ pint) orange juice
30 ml (2 tbsp) clear honey
75 g (3 oz) butter or margarine
175 g (6 oz) ginger biscuits, finely crushed
350 g (12 oz) full-fat soft cheese

150 ml (¼ pint) soured cream
15 ml (1 level tbsp) gelatine
clear honey, to glaze
toasted flaked almonds, to decorate
 (optional)

1 PUT THE APRICOTS and the orange juice in a medium bowl. Cover and
cook on HIGH for 5–7 minutes, until softened, stirring once.

2 RESERVE A FEW of the apricots for the decoration and drain on absorbant
kitchen paper. Put the remainder in a blender or food processor and purée
until smooth. Stir in the honey and leave to cool.

3 CUT THE BUTTER or margarine into small pieces, put in a medium bowl
and cook on HIGH for 1 minute or until melted. Stir in the crushed biscuits
and mix together.

4 PRESS THE MIXTURE into the base of a greased 20.5 cm (8 inch) loose-
bottom deep cake tin. Chill while finishing the filling.

5 BEAT THE CHEESE, cream and apricot purée together.

6 IN A SMALL bowl, sprinkle the gelatine over 60 ml (4 tbsp) water and
leave to soak for 1 minute. Cook on HIGH for 30–50 seconds until dissolved,
stirring frequently. Do not boil.

7 STIR THE GELATINE into the apricot mixture and pour on top of the biscuit
base. Level the surface, then chill for 3–4 hours until set.

8 TO SERVE, CAREFULLY remove the cheesecake from the tin and place on a
serving plate. Arrange the reserved apricots around the edge and brush
with a little honey to glaze. Sprinkle with a few almonds, if liked.

CHAPTER 2

STEWING AND BRAISING

*C*onventionally cooked stews and braises depend on long, slow cooking to tenderise tough cuts of meat and allow the flavours of the vegetables and herbs to combine. For this reason, there is little point in using the microwave for this type of stew. Instead, use it for stews made with tender cuts of meat, poultry, vegetables and any other ingredients which do not need tenderising.

1   Use only tender cuts of meat, poultry and vegetables for stewing and braising in the microwave. Thus because microwave cooking is not a slow cooking method so tougher cuts do not tenderise in the cooking time.

2   Stews cooked in the microwave require less added liquid because foods cook in their own moisture.

3   Cover the container with a large plate. Cling film is no longer recommended for use in the microwave.

4   Reposition braised dishes, when the meat is cooked on a bed of vegetables, to ensure even cooking.

5   To ensure even cooking, the ingredients of a stew should be stirred regularly throughout the cooking time.

CHICKEN WITH GINGER
SERVES 2 as a main course

▲

To minimize washing up, cook this well-flavoured stew in a serving dish—making sure that it is ovenproof—and serve from the dish.

15 ml (1 tbsp) vegetable oil
1 small onion, skinned and finely chopped
1 small garlic clove, skinned and crushed
1 cm ($\frac{1}{2}$ inch) piece fresh root ginger, peeled and finely grated
2.5 ml ($\frac{1}{2}$ level tsp) coriander seeds
2.5 ml ($\frac{1}{2}$ level tsp) cumin seeds

15 ml (1 level tbsp) tomato purée
225 g (8 oz) potato, scrubbed and cut into 2.5 cm (1 inch) pieces
2 chicken quarters, each about 225 g (8 oz), skinned
200 ml (7 fl oz) chicken stock
salt and pepper

1 PUT THE OIL, onion, garlic and ginger in a medium bowl. Cover and cook on HIGH for 5–7 minutes or until softened, stirring occasionally.

2 STIR IN THE coriander, cumin, tomato purée and potato and cook on HIGH for 2 minutes.

3 CUT EACH CHICKEN quarter into two pieces and add to the spice and potato mixture. Stir in the stock and salt and pepper to taste. Mix thoroughly.

4 RE-COVER AND COOK on HIGH for 25–30 minutes or until the chicken is tender, stirring occasionally. Serve hot with a green vegetable.

CHICKEN VERONIQUE
SERVES 4

▲

This rich and creamy dish is ideal for a dinner party.

50 g (2 oz) butter or margarine
50 g (2 oz) plain flour
300 ml ($\frac{1}{2}$ pint) chicken stock
300 ml ($\frac{1}{2}$ pint) dry white wine
450 g (1 lb) chicken breast fillets, skinned and cut into 5 cm (2 inch) pieces

150 ml ($\frac{1}{4}$ pint) single cream
175 g (6 oz) seedless green grapes
salt and pepper

1 PUT THE BUTTER or margarine, flour, stock and wine in a large bowl and whisk together. Cook on HIGH for 6–7 minutes, whisking every minute, until the sauce has boiled and thickened.

2 STIR IN THE chicken, cover and cook on HIGH for 6–7 minutes until the chicken is tender, stirring occasionally.

3 STIR IN THE cream and grapes and season to taste with salt and pepper. Re-cover and cook on LOW for 4–5 minutes until hot. Do not boil.

Chicken with Ginger

CHICKEN WITH TOMATOES AND OLIVES
―――――――――SERVES 2 as a main course―――――――――
▲

Serve this colourful dish with boiled rice to mop up the delicious sauce.

15 ml (1 tbsp) vegetable oil
1 medium onion, skinned and chopped
1 garlic clove, skinned and crushed
3 rashers smoked streaky bacon, rinded and
  chopped
1 green pepper, seeded and chopped
2 chicken thighs, skinned
2 chicken drumsticks, skinned
397 g (14 oz) can tomatoes, drained and
  chopped

10 ml (2 level tsp) tomato purée
5 ml (1 level tsp) paprika
pinch of sugar
salt and pepper
50 g (2 oz) black olives
30 ml (2 tbsp) chopped fresh parsley
chopped fresh parsley, to garnish

1 PUT THE OIL, onion, garlic, bacon and green pepper in a large bowl, cover and cook on HIGH for 5 minutes or until the vegetables have softened, stirring occasionally.

2 ADD THE CHICKEN, tomatoes, tomato purée, paprika, sugar and salt and pepper to taste. Re-cover and cook on HIGH for 20 minutes or until the chicken is tender, turning the chicken over once during cooking and stirring occasionally.

3 STIR IN THE olives and parsley and cook, uncovered, on HIGH for 5 minutes, stirring once. Garnish with chopped parsley and serve hot.

BRAISED CHICKEN IN ANCHOVY SAUCE
―――――――――SERVES 4―――――――――
▲

This recipe is based on an Italian recipe which adapts well to cooking in a microwave. Canned anchovies are salty so do not add salt when seasoning.

1 small onion, skinned and finely chopped
1 garlic clove, skinned and finely chopped
15 ml (1 tbsp) vegetable oil
60 ml (4 tbsp) dry white wine
2.5 ml ($\frac{1}{2}$ tsp) dried oregano
pepper

4 chicken breasts, skinned
150 ml ($\frac{1}{4}$ pint) chicken stock
10 ml (2 level tsp) plain flour
4 anchovy fillets, mashed
black olives, to garnish

1 PUT THE ONION, garlic and oil in a large shallow dish and cook on HIGH for 4–5 minutes until the onion has softened.

2 ADD THE WINE and cook on HIGH for 2 minutes until boiling. Add the oregano and season to taste with pepper.

3 ARRANGE THE CHICKEN on top of the onion and pour in the stock. Cover and cook on HIGH for 7–9 minutes until the chicken is tender, turning the chicken once during cooking.

4 ARRANGE THE CHICKEN on a warmed serving dish. Whisk the flour and anchovies into the dish and cook on HIGH for 5 minutes until the sauce is boiling and thickened slightly, whisking frequently.

5 POUR THE SAUCE over the chicken and serve, garnished with black olives.

TURKEY IN SPICED YOGURT
————————SERVES 2 as a main course————————
▲

If you cannot find turkey, chicken fillet works equally well. Marinating the meat overnight before cooking greatly improves the flavour, but if time is short marinate for 2–3 hours instead.

5 ml (1 level tsp) ground cumin
5 ml (1 level tsp) ground coriander
5 ml (1 level tsp) ground turmeric
2.5 ml ($\frac{1}{2}$ level tsp) ground cardamom
150 ml ($\frac{1}{4}$ pint) natural yogurt
salt and pepper
350 g (12 oz) boneless turkey, skinned

15 ml (1 tbsp) vegetable oil
1 small onion, skinned and sliced
30 ml (2 level tbsp) desiccated coconut
15 ml (1 level tbsp) plain flour
15 ml (1 level tbsp) mango chutney
75 ml (3 fl oz) chicken stock
chopped fresh parsley, to garnish

1 IN A MEDIUM bowl, mix the spices, yogurt, salt and pepper together. Cut the turkey into 2.5 cm (1 inch) cubes and stir into the spiced yogurt. Cover and leave in the refrigerator to marinate overnight.

2 THE NEXT DAY, put the oil and onion in a medium bowl, cover and cook on HIGH for 3–4 minutes or until softened.

3 STIR IN THE coconut and flour and cook on HIGH for 30 seconds, then gradually stir in the chutney, chicken stock and the turkey and yogurt mixture.

4 RE-COVER AND COOK on HIGH for 5–6 minutes or until the meat is tender, stirring occasionally.

5 LEAVE TO STAND for 5 minutes. Adjust the seasoning if necessary, then turn into a warmed serving dish, garnish with parsley and serve immediately.

Below: Cinnamon Lamb with Almonds and Apricots (page 38)
Opposite: Vegetable Goulash (page 43)

CINNAMON LAMB WITH ALMONDS AND APRICOTS

―――――――――――SERVES 2 as a main course―――――――

▲

Here, a browning dish acts like a conventional frying pan to brown and seal the meat before it is stewed on a LOW microwave setting.

25 g (1 oz) whole blanched almonds
50 g (2 oz) dried apricots, halved
350 g (12 oz) lamb fillet
15 ml (1 level tbsp) plain flour
10 ml (2 level tsp) ground cinnamon
2.5 ml ($\frac{1}{2}$ level tsp) ground cumin

salt and pepper
15 ml (1 tbsp) vegetable oil
75 ml (3 fl oz) chicken stock
1 bay leaf
30 ml (2 tbsp) natural yogurt

1 PUT THE ALMONDS on a large flat plate and cook on HIGH for 6 minutes or until lightly browned, stirring occasionally. Set aside.

2 PUT THE APRICOTS in a small bowl with 150 ml ($\frac{1}{4}$ pint) water. Cover and cook on HIGH for 5 minutes. Leave to stand.

3 HEAT A BROWNING dish for 8–10 minutes, or according to the manufacturer's instructions.

4 MEANWHILE, CUT THE lamb into 2.5 cm (1 inch) slices and flatten slightly with a meat mallet or a rolling pin. Cut each slice in half.

5 MIX THE FLOUR, cinnamon, cumin, salt and pepper together and use to coat the meat.

6 ADD THE OIL to the browning dish, then quickly stir in the meat. Cook on HIGH for 2 minutes, then turn the meat over and cook on HIGH for a further 2 minutes.

7 STIR IN THE stock and bay leaf and mix well together. Cover and cook on LOW for 10 minutes or until the meat is tender, stirring occasionally.

8 DRAIN THE APRICOTS and stir into the dish. Cook on HIGH for 2–3 minutes or until the apricots are hot. Stir in the yogurt and more seasoning if necessary. Serve hot, sprinkled with the toasted almonds.

MINTED LAMB MEAT BALLS

—————————SERVES 4 as a main course—————————

▲

Meat balls braise very quickly in the microwave, but must be rearranged during cooking to ensure that they cook evenly.

225 g (8 oz) crisp green cabbage, roughly
    chopped
1 medium onion, skinned and quartered
450 g (1 lb) lean minced lamb
2.5 ml ($\frac{1}{2}$ level tsp) ground allspice
salt and pepper

397 g (14 oz) can tomato juice
1 bay leaf
10 ml (2 tsp) chopped fresh mint or 2.5 ml
    ($\frac{1}{2}$ level tsp) dried
15 ml (1 tbsp) chopped fresh parsley

1 PUT THE CABBAGE and onion in a blender or food processor and process until finely chopped. Transfer to a large bowl, cover and cook on HIGH for 2–3 minutes until the vegetables are softened. Leave for 5 minutes to cool.

2 ADD THE LAMB and allspice and season with salt and pepper. Beat well.

3 USING WET HANDS, shape the lamb mixture into 16 small balls and place them in a single layer in a shallow dish. Cook on HIGH for 5 minutes, carefully turning and re-positioning the meat balls after 3 minutes.

4 MIX THE TOMATO juice with the bay leaf, mint and parsley and pour over the meat balls. Cover and cook on HIGH for 5–6 minutes until the sauce is boiling and the meat balls are cooked. Allow them to stand for 5 minutes, then skim off any fat. Serve with rice or noodles.

Below: Mushroom, Courgette and Bean Stew (page 42), Herb, Cheese and Olive Bread (page 180)
Opposite: Malaysian Prawns (page 44)

MARINATED PORK WITH PEANUTS

————————SERVES 2 as a main course————————

▲

Marinating the chops before cooking helps to tenderize the meat and allows the flavours to permeate all the way through. Thirty minutes will do—but if you have more tiime, 2–3 hours is better.

60 ml (4 level tbsp) crunchy peanut butter
30 ml (2 tbsp) soy sauce
10 ml (2 tsp) lemon juice
2.5 ml (½ level tsp) mild chilli powder
5 ml (1 level tsp) ground cumin

10 ml (2 level tsp) dark soft brown sugar
1 large garlic clove, skinned and crushed
2 pork loin chops, about 2.5 cm (1 inch) thick

1 PUT THE PEANUT butter in a small bowl and gradually mix in the soy sauce, lemon juice, chilli powder, cumin, sugar and garlic.

2 TRIM ANY EXCESS fat from the pork and prick all over with a fork. Coat with the peanut mixture.

3 PUT IN A shallow dish, cover and leave in the refrigerator to marinate for at least 30 minutes.

4 To COOK THE pork, cover and cook on HIGH for 7–9 minutes or until tender, turning once during cooking. Serve hot with boiled rice, if liked.

MUSHROOM, COURGETTE AND BEAN STEW

————————SERVES 4 as a main course————————

▲

This makes a colourful and delicious vegetarian meal. Include button and flat mushrooms for the best flavour and mixed fresh herbs of your choice. When preparing mushrooms, do not be tempted to wash them as they absorb water quickly and lose their flavour and texture— simply wipe with a damp cloth or kitchen paper. Likewise they should never be peeled as most of the flavour is in the skin.

25 g (1 oz) butter or margarine
1 medium onion, skinned and chopped
25 g (1 oz) wholemeal flour
450 ml (¾ pint) vegetable stock
15 ml (1 level tbsp) mild wholegrain mustard
450 g (1 lb) cooked beans such as flageolet, borlotti, or black-eye beans (see page 218), or two 425 g (15 oz) cans beans, drained and rinsed

225 g (8 oz) mushrooms
450 g (1 lb) courgettes
45 ml (3 tbsp) chopped fresh mixed herbs
salt and pepper

1 PUT THE BUTTER or margarine and the onion in a large bowl. Cover and cook on HIGH for 2–3 minutes or until slightly softened. Stir in the flour and cook on HIGH for 1 minute, then gradually stir in the stock.

2 COOK ON HIGH for 4–5 minutes or until boiling and thickened, stirring frequently.

3 ADD THE MUSTARD, beans and the mushrooms (cut in half if large) and cook on HIGH for 2–3 minutes.

4 MEANWHILE, CUT THE courgettes into 1 cm ($\frac{1}{2}$ inch) slices. Stir the courgettes and half of the herbs into the stew. Cover and cook on HIGH for 5–6 minutes or until the courgettes are just cooked. Season to taste with salt and pepper and stir in the remaining herbs. Serve with brown rice or Herb, Cheese and Olive Bread (see page 180).

VEGETABLE GOULASH
SERVES 2 as a main course

▲

A warming winter vegetable stew to serve topped with grated cheese, accompanied by lots of crusty, wholemeal bread.

30 ml (2 tbsp) vegetable oil
1 medium onion, skinned and chopped
1 green pepper, seeded and chopped
15 ml (1 level tbsp) sweet paprika
2.5 ml ($\frac{1}{2}$ level tsp) caraway seeds
45 ml (3 level tbsp) medium oatmeal
450 ml ($\frac{3}{4}$ pint) tomato juice
2 medium carrots, cut into 0.5 cm ($\frac{1}{4}$ inch) slices

2 medium courgettes, cut into 2.5 cm (1 inch) slices
100 g (4 oz) button mushrooms
freshly grated nutmeg
salt and pepper
30 ml (2 tbsp) soured cream or natural yogurt
parsley sprigs, to garnish

1 PUT THE OIL, onion and pepper in a large bowl. Cover and cook on HIGH for 5–7 minutes or until softened, stirring occasionally.

2 STIR IN THE paprika and caraway seeds and cook on HIGH for 1 minute. Stir in the oatmeal and gradually stir in the tomato juice.

3 STIR THE CARROTS, courgettes and mushrooms into the paprika mixture and mix well. Season to taste with nutmeg and salt and pepper.

4 RE-COVER AND COOK on HIGH for 15–20 minutes or until the vegetables are tender. Serve with the soured cream or yogurt spooned on top, garnished with parsley sprigs.

RATATOUILLE
—SERVES 4 as a main course—

▲

This classic Provençal stew can be served hot on its own as a main dish or cold as an accompaniment.

75 ml (3 tbsp) olive oil
450 g (1 lb) onions, skinned and thinly sliced
1 garlic clove, skinned and crushed
450 g (1 lb) tomatoes, skinned, seeded and chopped, or 397 g (14 oz) can tomatoes with their juice

2 red or green peppers, seeded and sliced
450 g (1 lb) courgettes, sliced
450 g (1 lb) aubergines, thinly sliced
30 ml (2 level tbsp) tomato purée
salt and pepper
bouquet garni

1 PUT THE OIL in a large bowl with the onions and the garlic. Cover and cook on HIGH for 5–7 minutes until soft.

2 ADD THE REMAINING ingredients, cover and cook on HIGH for 25–30 minutes. The vegetables should be soft and well mixed but retain their shape and most of the liquid should have evaporated. Serve hot or cold.

MALAYSIAN PRAWNS
—SERVES 4 as a main course or 6 as a starter—

▲

These prawns taste delicious when served with grilled poppadums. The sauce can be prepared in advance, and the prawns added and heated through just before serving.

1 small onion, skinned and finely chopped
1 garlic clove, skinned and chopped
3 large tomatoes, roughly chopped
2.5 cm (1 inch) piece fresh root ginger, peeled and crushed
2.5 ml ($\frac{1}{2}$ level tsp) ground turmeric
2.5 ml ($\frac{1}{2}$ level tsp) ground cumin

15 ml (1 tbsp) red wine vinegar
5 ml (1 level tsp) tomato purée
50 g (2 oz) creamed coconut, crumbled
450 g (1 lb) cooked peeled prawns
salt and pepper
chopped spring onions, to garnish

1 PUT THE ONION, garlic, tomatoes, ginger, turmeric, cumin, vinegar, tomato purée, coconut and 150 ml ($\frac{1}{4}$ pint) boiling water in a medium bowl. Cook on HIGH for 8 minutes until thickened, stirring occasionally.

2 ADD THE PRAWNS and stir together. Cover and cook on HIGH for 2–3 minutes until the prawns are heated through, stirring once. Season to taste with salt and pepper. Garnish with spring onions and serve hot.

CHAPTER 3

POACHING AND STEAMING

*F*ish, chicken vegetables and fruits are all suitable for poaching or steaming in a microwave. The foods are cooked in very little liquid, usually much less than when cooked conventionally, because the food cooks in its own moisture. If a dish is covered, the steam which is trapped by the lid cooks the food.

1   All foods should be of an even size since this ensures that the food is cooked evenly for the same length of time.

2   Poached and steamed foods need only a small amount of liquid added to the dish because microwaving is a naturally moist form of cooking.

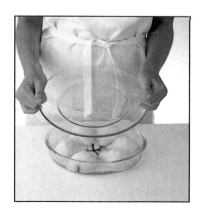

3   Cover the container with a large plate. Cling film is no longer recommended for use in the microwave.

4   Foods that can be stirred, such as vegetables, should be stirred during cooking to ensure even cooking.

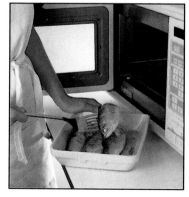

5   Whole foods, that cannot be stirred, should be repositioned to ensure even cooking.

SMOKED HADDOCK MOUSSES

————————SERVES 6 as a starter————————
▲

*Smoked fish has a high salt content so do not add extra salt. These
low-calorie mousses are served chilled.*

350 g (12 oz) smoked haddock fillet
100 g (4 oz) cottage or curd cheese
150 ml ($\frac{1}{4}$ pint) natural yogurt
grated rind and juice of $\frac{1}{2}$ lemon

15 ml (1 tbsp) chopped fresh parsley
pepper
5 ml (1 level tsp) gelatine
lemon slices, to garnish

1 PUT THE HADDOCK in a shallow dish with 30 ml (2 tbsp) water. Cover and
cook on HIGH for 3 minutes or until the fish is tender.

2 DRAIN AND FLAKE the fish, discarding the skin and bones, and put in a
blender or food processor. Add the cheese, yogurt, lemon rind, parsley and
pepper to taste and work until smooth.

3 PUT THE LEMON juice in a small bowl and sprinkle in the gelatine. Cook
on LOW for 1–1$\frac{1}{2}$ minutes or until the gelatine has dissolved, stirring
occasionally. Add to the fish mixture and mix well together.

4 DIVIDE THE FISH mixture equally between six individual ramekin dishes.
Chill for at least 1 hour before serving.

5 IF LIKED, TURN the mousses out on to individual plates. Garnish with
lemon slices and serve with toasted wholemeal bread.

POACHED TROUT WITH MUSHROOM SAUCE

————————SERVES 2 as a main course————————
▲

*Whole fish cook beautifully in the microwave, but it is important to
slash their skins to prevent them from bursting during cooking.*

2 whole trout, each weighing about 350 g
    (12 oz), cleaned
4 spring onions, trimmed and thinly sliced
100 g (4 oz) mushrooms, sliced
sprig of parsley

5 ml (1 tsp) lemon juice
salt and pepper
90 ml (6 tbsp) hot fish or vegetable stock
30 ml (2 tbsp) single cream
lemon wedges, to garnish

1 SLASH THE SKIN of the fish in two or three places on both sides, then
arrange them in an oblong dish, with their backbones to the edges of the
dish. Add the spring onions, mushrooms, parsley, lemon juice and season
to taste with salt and pepper. Pour over the stock.

2 COVER AND COOK on HIGH for 7 minutes or until tender.

3 ARRANGE THE FISH on warmed serving dishes. Stir the cream into the mushroom mixture and cook on HIGH for a further 3 minutes or until reduced slightly. Garnish the fish with lemon wedges and serve the sauce separately.

POTATO-TOPPED FISH PIE
————————SERVES 4 as a main course————————
▲

This is a basic recipe for fish pie which can be varied with the addition of peeled prawns, mushrooms, grated cheese, or fresh herbs.

4 medium potatoes, each weighing about 175 g (6 oz)
50 g (2 oz) butter or margarine
450 ml (¾ pint) milk
salt and pepper
450 g (1 lb) white fish fillets, such as cod, haddock or coley

225 g (8 oz) smoked haddock
25 g (1 oz) plain flour
freshly grated nutmeg
2 eggs, hard-boiled and shelled
45 ml (3 tbsp) chopped fresh parsley

1 SCRUB THE POTATOES and prick all over with a fork. Arrange on absorbent kitchen paper in a circle in the cooker and cook on HIGH for 12–14 minutes or until tender, turning over once.

2 WHEN THE POTATOES are cooked, remove from the cooker and set aside to cool slightly. Peel the potatoes and put in a bowl. Add 25 g (1 oz) of the butter or margarine and about 75 ml (5 tbsp) of the milk, or enough to make a soft mashed potato. Mash thoroughly together. Season to taste with salt and pepper.

3 PUT THE FISH in a single layer in a large shallow dish with 30 ml (2 tbsp) of the milk. Cover and cook on HIGH for 4–5 minutes or until the fish flakes easily.

4 REMOVE THE FISH from the cooker. Strain the cooking liquid from the fish into a medium bowl and reserve. Remove and discard the skin and any bones from the fish. Flake the fish and put in a flameproof serving dish.

5 PUT THE REMAINING milk, butter or margarine and flour into the bowl with the reserved cooking liquid and cook on HIGH for 4–5 minutes or until boiling and thickened, whisking every minute. Season to taste with salt, pepper and nutmeg. Roughly chop the hard-boiled eggs and stir into the sauce with the parsley. Pour over the fish.

6 SPOON OR PIPE the mashed potato on top of the fish mixture. Cook on HIGH for 4–5 minutes or until the pie is hot, then brown under a hot grill if liked. Serve at once with a green vegetable.

SOLE AND SPINACH ROULADES
————————SERVES 4 as a main course————————
▲

Serve three of these pretty spinach-filled roulades per person.

12 sole fillets, each weighing about 75 g (3 oz), skinned
5 ml (1 level tsp) fennel seeds, crushed
salt and pepper
12 spinach or sorrel leaves, washed

15 ml (1 tbsp) dry white wine
45 ml (3 tbsp) Greek strained yogurt
pinch of ground turmeric
spinach or sorrel leaves, to garnish

1 PLACE THE SOLE fillets, skinned side up, on a chopping board. Sprinkle with the fennel seeds and season to taste with salt and pepper. Lay a spinach or sorrel leaf, vein side up on top of each fillet, then roll up and secure with a wooden cocktail stick.

2 ARRANGE THE FISH in a circle around the edge of a large shallow dish and pour over the wine. Cover and cook on HIGH for 6–7 minutes until tender.

3 USING A SLOTTED spoon, transfer the fish to a serving plate.

4 GRADUALLY STIR THE yogurt and turmeric into the cooking liquid. Season to taste with salt and pepper and cook on HIGH for 1–2 minutes until slightly thickened, stirring occasionally. Serve the roulades with a little of the sauce poured over, garnished with spinach or sorrel leaves.

POACHED PLAICE IN VERMOUTH
————————SERVES 2 as a main course————————
▲

Vermouth is a fortified wine flavoured with herbs.

15 g (½ oz) butter or margarine
2 spring onions, trimmed and sliced
30 ml (2 tbsp) dry vermouth
15 ml (1 tbsp) lemon juice
1 small garlic clove, skinned and crushed
pinch of dill

a few green peppercorns, crushed
salt and pepper
2 plaice fillets, each weighing 175 g (6 oz), skinned
lemon slices, to garnish

1 USE THE BUTTER or margarine to grease a shallow dish that will just hold the fish. Put the spring onions in the dish with the vermouth, lemon juice, garlic, dill, peppercorns and salt and pepper to taste. Add the fish to this mixture and leave to marinate for 10 minutes, turning once.

2 COVER AND COOK on HIGH for 4 minutes until tender. Garnish with lemon slices. Serve with creamed potatoes and spinach with nutmeg.

Sole and Spinach Roulades

FISH BALLS IN A WALNUT SAUCE

SERVES 4 as a main course

▲

This is an unusual way of serving bland fish. Be careful when rearranging the fish balls in step 5 or they will break up. Serve the fish balls with a rice pilaff.

450 g (1 lb) white fish fillets, such as haddock or whiting
30 ml (2 tbsp) milk
50 g (2 oz) fresh brown breadcrumbs
1 small onion, skinned and very finely chopped.
30 ml (2 tbsp) chopped fresh coriander or parsley
salt and pepper
1 egg yolk

100 g (4 oz) walnut halves
2 garlic cloves, skinned and crushed
5 ml (1 level tsp) paprika
5 ml (1 level tsp) ground coriander
pinch of ground cloves
30 ml (2 tbsp) white wine vinegar
450 ml (¾ pint) fish or vegetable stock
walnut halves or coriander or parsley sprigs, to garnish

1 PUT THE FISH in a shallow dish with the milk. Cover and cook on HIGH for 4–5 minutes until the fish flakes easily. Flake the fish, discarding the skin and any bones. Reserve the cooking liquid.

2 PUT THE FISH, breadcrumbs, onion, fresh coriander or parsley and salt and pepper to taste in a blender or food processor and purée until smooth. Gradually add the egg yolk to bind the mixture together to make a fairly stiff consistency.

3 SHAPE THE MIXTURE into 20 walnut-size balls. Chill while making the sauce.

4 PUT THE WALNUTS, garlic, paprika, ground coriander, cloves and vinegar in the rinsed-out bowl of the blender or food processor and purée until smooth. Transfer to a large bowl and cook on HIGH for 2 minutes, stirring frequently.

5 ADD THE RESERVED cooking liquid and the stock to the walnut mixture and cook on HIGH for 5–6 minutes or until boiling, stirring once. Carefully add the fish balls and cook on MEDIUM for 5–6 minutes or until they feel firm to the touch, rearranging them carefully once during cooking. Garnish with walnut halves or coriander or parsley sprigs and serve hot with Herby Rice Pilaff (see page 24), if liked, or a selection of vegetables.

Fish Balls in a Walnut Sauce

SCALLOPS WITH PESTO SAUCE
————————————SERVES 2 as a main course————————————
▲

Serve this delicious dish as a special dinner for two. The pesto sauce can be made earlier in the day and just needs stirring gently before serving.

15 g (½ oz) fresh basil leaves
1 garlic clove, skinned
15 ml (1 level tbsp) pine nuts
25 ml (1 fl oz) olive oil
salt and pepper

15 g (½ oz) Parmesan cheese, grated
30 ml (2 tbsp) double cream
6 large shelled scallops, weighing about
   400 g (14 oz)
fresh basil, to garnish

1 To MAKE THE pesto sauce, put the basil, garlic, pine nuts, oil, salt and pepper in a blender or food processor and process until smooth. Fold in the cheese and cream.

2 CUT THE CORALS from the scallops and set aside. Slice the white part across into two discs.

3 ARRANGE THE WHITE parts in a circle in a large shallow dish. Cover and cook on HIGH for 2 minutes or until the scallops are just opaque.

4 ADD THE RESERVED corals and cook on HIGH for a further 1 minute until the corals are tender.

5 DRAIN THE SCALLOPS, arrange on two plates and spoon over the pesto sauce. Garnish with fresh basil and serve immediately with French bread and a salad, if liked.

FISH STEAKS WITH HAZELNUT SAUCE
————————————SERVES 4 as a main course————————————
▲

Do not overcook the hazelnuts or the sauce will have a bitter, burnt flavour.

100 g (4 oz) hazelnuts
4 white fish steaks, such as haddock or cod,
   each weighing about 175 g (6 oz)
45 ml (3 tbsp) dry white wine, fish or
   vegetable stock

150 ml (¼ pint) double cream or Greek
   strained yogurt
salt and pepper
ground mace
watercress sprigs, to garnish

1 SPREAD THE HAZELNUTS out on a large plate and cook on HIGH for 30 seconds. Tip on to a clean tea towel and rub off the loose brown skin. Return the nuts to the cooker and cook on HIGH for a further 6–10 minutes, stirring frequently until lightly browned, then chop finely.

2 ARRANGE THE FISH in a single layer in a large shallow dish and pour over the wine or stock. Cover and cook on HIGH for 6–7 minutes or until the fish is cooked. Transfer the fish to a serving dish.

3 ADD THE HAZELNUTS and cream or yogurt to the cooking dish and cook on HIGH for 2 minutes until hot. Season to taste with salt, pepper and mace, then pour over the fish. Garnish with watercress sprigs and serve immediately.

GREY MULLET STUFFED WITH GARLIC AND HERBS
————————SERVES 2–3 as a main course————————
▲

Grey mullet is a round fish which looks and tastes similar to sea bass. Here it is complemented by fresh mixed herbs and garlic.

50 g (2 oz) hazelnuts
1 grey mullet, weighing about 700 g (1½ lb), cleaned and scaled
3 garlic cloves, skinned
finely grated rind and juice of 1 lemon
45 ml (3 tbsp) chopped fresh mixed herbs, such as parsley, basil, tarragon, chervil, mint, coriander

salt and pepper
15 ml (1 tbsp) olive oil
fresh herbs, to garnish

1 SPREAD OUT THE hazelnuts on an ovenproof plate. Cook on HIGH for 3–4 minutes until lightly toasted. Set aside to cool.

2 USING A SHARP knife, slash the fish three or four times on each side.

3 PUT THE HAZELNUTS, garlic, lemon rind, half of the lemon juice and the herbs in a blender or food processor and work until a coarse paste. Season to taste with salt and pepper. Spoon a little of the paste into the slashes and use the rest to stuff the fish.

4 PLACE THE FISH on a large ovenproof serving plate. Mix the remaining lemon juice with the oil and spoon over the fish. Season generously with pepper.

5 COVER AND COOK on HIGH for 6–7 minutes or until tender. Serve garnished with fresh herbs.

MUSSELS IN WHITE WINE
————————SERVES 2 as a light meal————————
▲

*Mussels need careful preparation, but they are easy to cook in the
microwave and remain succulent and tender.*

900 g (2 lb) fresh mussels
1 small onion, skinned and finely chopped
1 garlic clove, skinned and crushed
75 ml (5 tbsp) dry white wine

75 ml (5 tbsp) fish stock or water
30 ml (2 tbsp) chopped fresh parsley
salt and pepper

1 TO CLEAN THE mussels, put them in a sink or bowl and scrub thoroughly
with a hard brush. Wash them in several changes of water.

2 SCRAPE OFF ANY 'beards' or tufts protruding from the shells. Discard any
damaged mussels or any that do not close when tapped with a knife.

3 PUT THE ONION, garlic, wine, stock and mussels in a large bowl. Cover
and cook on HIGH for 3–5 minutes or until all the mussels have opened,
removing the mussels on the top as they open and shaking the bowl
occasionally. Discard any mussels that have not opened.

4 PILE THE MUSSELS in a warmed serving dish. Stir the parsley into the
liquid remaining in the bowl and season to taste with salt and pepper. Pour
over the mussels and serve immediately with lots of crusty bread.

SALMON EN PAPILLOTE
————————SERVES 2 as a main course————————
▲

*Cooking fish en papillote ensures that all the flavour and juice are
retained inside a greaseproof paper parcel.*

2 salmon fillets, each weighing 150 g (5 oz)
butter
salt and pepper

100 g (4 oz) cooked peeled prawns
lemon juice

1 PLACE THE SALMON fillets on squares of buttered greaseproof paper and
season to taste with salt and pepper.

2 SPOON THE PRAWNS over the fish and dot with butter. Squeeze lemon
juice over each.

3 FOLD THE PAPER edges tightly to make a plump parcel. Place the parcels
on an ovenproof plate and cook on HIGH for $3\frac{1}{2}$–4 minutes until the fish is
tender. Serve with Hollandaise Sauce, if liked (see page 143).

Mussels in White Wine

LEAF-WRAPPED MULLET
SERVES 4 as a main course

▲

Swiss chard makes a good wrapping for fish as the leaves are large and pliable and retain their bright green colour when cooked. It is a member of the beetroot family and is also known as seakale beet or silver beet.

4 red mullet, each weighing about 275 g (10 oz), cleaned and scaled
3–4 garlic cloves, skinned
45 ml (3 tbsp) olive oil

30 ml (2 tbsp) white wine vinegar
salt and pepper
8 large Swiss chard or spinach leaves, trimmed

1 USING A SHARP knife, slash the mullet three times on each side. Roughly chop the garlic and sprinkle into the slashes. Whisk the oil and vinegar together and season to taste with salt and pepper.

2 PUT THE FISH in a shallow dish and pour over the oil and vinegar. Leave in a cool place for 30 minutes to marinate.

3 REMOVE THE FISH from the marinade and wrap each of them in two of the chard or spinach leaves. Return the wrapped fish to the dish containing the marinade.

4 COVER AND COOK on HIGH for 6–8 minutes or until the fish is tender, rearranging once and basting with the marinade during cooking.

5 SERVE THE FISH in their leaf parcels, with a little of the marinade spooned over.

CHICKEN WITH APPLE AND CURRIED MAYONNAISE
SERVES 2 as a light meal

▲

This is a variation of the well-known Coronation chicken recipe. It is delicious served chilled on a summer's day.

1 medium onion, skinned and chopped
2 chicken breast fillets, skinned
60 ml (4 tbsp) dry white wine
1 bay leaf
large pinch of dried mixed herbs
20 ml (4 level tsp) mild curry powder
1 small green pepper

1 small red apple
10 ml (2 tsp) lemon juice
90 ml (6 tbsp) mayonnaise
15 ml (1 tbsp) apricot jam
salt and pepper
few lettuce leaves, to serve

1 PUT THE ONION, chicken breasts, wine, bay leaf and mixed herbs in a shallow dish. Cover and cook on HIGH for 5–6 minutes or until tender.

2 CUT THE CHICKEN into bite-sized pieces and set aside. Stir the curry powder into the cooking liquid and cook on HIGH for 2–3 minutes or until slightly reduced.

3 MEANWHILE, CORE, SEED and dice the green pepper and core and thinly slice the apple. Mix with the chicken, stir in the lemon juice and set aside.

4 STIR THE MAYONNAISE and apricot jam into the cooking liquid and season with salt and pepper. Pour over the chicken mixture and mix together until thoroughly coated.

5 LINE A SERVING dish with the lettuce leaves and pile in the chicken mayonnaise. Chill for about 20 minutes before serving.

CHICKEN BREASTS WITH GRUYÈRE CHEESE
SERVES 2 as a main course

▲

Although you could use another cheese, such as Cheddar, instead of the Gruyère cheese, the flavour won't be the same. The cheese should just be allowed to melt, which is why it is added at the end of the cooking. If overcooked it will become tough and rubbery.

200 ml (7 fl oz) milk
$\frac{1}{2}$ small onion, skinned
$\frac{1}{2}$ small carrot, sliced
$\frac{1}{2}$ celery stick, sliced
1 bay leaf
2 black peppercorns

15 g ($\frac{1}{2}$ oz) butter or margarine
15 g ($\frac{1}{2}$ oz) plain flour
salt and pepper
2 chicken breast fillets, skinned
50 g (2 oz) Gruyère cheese, thinly sliced
chopped fresh parsley, to garnish

1 PUT THE MILK, onion, carrot, celery, bay leaf and peppercorns in a medium bowl and cook on HIGH for 3 minutes or until boiling. Leave to infuse for 30 minutes.

2 STRAIN THE MILK and return it to the bowl, discarding the flavourings. Add the butter and flour and cook on HIGH for 2–3 minutes or until boiling and thickened, whisking frequently. Season to taste with salt and pepper.

3 CUT EACH CHICKEN breast in half widthways and put in two individual gratin dishes. Pour over the sauce. Cover and cook on HIGH for 4–5 minutes or until the chicken is tender.

4 PLACE THE SLICED cheese on top and cook on HIGH for 1 minute or until melted. Brown under a hot grill if liked. Garnish with parsley and serve hot, straight from the dish.

Below: Leaf-wrapped Mullet (page 56), Hake and Lime Kebabs (page 81)
Opposite: Chicken with Apple and Curried Mayonnaise (page 56)

CHEESE AND NUT STUFFED COURGETTES

-------------------SERVES 2 as a light meal-------------------

▲

Stuffed courgettes make a delicious light meal or can be served as a vegetable accompaniment.

4 medium courgettes
15 ml (1 tbsp) vegetable oil
1 garlic clove, skinned and crushed
15 ml (1 tbsp) chopped fresh parsley
1 small onion, skinned and finely chopped

50 g (2 oz) button mushrooms, chopped
salt and pepper
25 g (1 oz) walnuts, finely chopped
50 g (2 oz) Cheddar cheese, grated

1 TRIM THE COURGETTES and cut in half lengthways. Arrange in a single layer in a shallow dish.

2 POUR OVER 60 ML (4 tbsp) water, cover and cook on HIGH for 6–8 minutes or until just cooked.

3 DRAIN THE COURGETTES, reserving the cooking liquid. With a teaspoon, scoop out the flesh into a bowl, leaving a thin shell. Mash the flesh, pour off any excess liquid.

4 ADD THE OIL, garlic, parsley, onion, mushrooms, salt and pepper to the courgette flesh. Cover and cook on HIGH for 5 minutes until the vegetables are softened, stirring occasionally.

5 STIR IN THE walnuts and cheese. Spoon the mixture into the reserved shells and arrange in a shallow dish.

6 COOK ON HIGH for 1–2 minutes until hot. Serve with a green salad.

MANGE-TOUT IN CREAM DRESSING

-------------------SERVES 2 as an accompaniment-------------------

▲

Mange-tout are the perfect microwave vegetable; they need virtually no preparation and they cook in a flash, retaining all their colour.

175 g (6 oz) mange-tout, trimmed
60 ml (4 tbsp) single cream
finely grated rind of 1 small lemon
5 ml (1 tsp) lemon juice

salt and pepper
large pinch of soft brown sugar
large pinch of ground turmeric

1 PUT THE MANGE-TOUT into a medium bowl with 15 ml (1 tbsp) water. Cover and cook on HIGH for 3–4 minutes or until just tender, stirring once.

2 DRAIN AND RETURN to the bowl with the remaining ingredients. Cook on HIGH for 1–2 minutes to heat through, stirring once. Serve immediately.

OKRA WITH BABY ONIONS AND CORIANDER

————————SERVES 4–6 as an accompaniment————————

▲

Okra are green ribbed pods with white flesh and edible seeds. They are often referred to as 'ladies' fingers' because of their long tapering shape. When trimming the ends before cooking, take care not to cut into the flesh or a sticky substance will be released during cooking.

15 ml (1 tbsp) olive oil
15 ml (1 tbsp) coriander seeds, crushed
1 garlic clove, skinned and crushed
225 g (8 oz) baby onions, skinned and
  halved

450 g (1 lb) okra, trimmed
60 ml (4 tbsp) vegetable stock
salt and pepper

1 PUT THE OIL, coriander and garlic in an ovenproof serving bowl. Cook on HIGH for 2 minutes, stirring once.

2 ADD THE ONIONS, okra and stock and mix well together. Cover and cook on HIGH for 5–7 minutes until the onions and okra are tender, stirring occasionally. Season with salt and pepper to taste and serve hot.

CARROTS WITH ORANGE SEGMENTS

————————SERVES 2 as an accompaniment————————

▲

Carrots are naturally sweet and combine beautifully with oranges.

2 large oranges
150 g (5 oz) carrots, trimmed and scrubbed
15 g ($\frac{1}{2}$ oz) butter
10 ml (2 tsp) lemon juice

30 ml (2 tbsp) clear honey
15 ml (1 tbsp) chopped fresh parsley
salt and pepper

1 THINLY PARE THE rind from half of one of the oranges and cut into very thin strips. Put the strips in a medium bowl and squeeze in the juice from the orange.

2 REMOVE THE RIND and pith from the second orange, cut the orange into segments and set aside.

3 CUT THE CARROTS into fingers 1 cm ($\frac{1}{2}$ inch) wide and 5 cm (2 inches) long. Stir them into the bowl with the butter. Cover and cook on HIGH for 5–7 minutes or until just tender, stirring occasionally.

4 UNCOVER AND STIR in the orange segments, lemon juice, honey and parsley. Season to taste with salt and pepper. Cook on HIGH for 1–1$\frac{1}{2}$ minutes or until just heated through, then serve immediately.

Below: Okra, Baby Onions and Coriander (page 61), Baby Carrots, Watercress and Orange (page 64)
Opposite: Blackcurrant Jelly with Fresh Fruit (page 66)

PETITS POIS À LA FRANCAISE
————————SERVES 2 as an accompaniment————————
▲

Shelling peas takes time but there is something very satisfying about this task, especially if it can be done sitting in the garden on a sunny day—and the result is well worth the effort.

1 lettuce heart	6 spring onions, trimmed and sliced
15 g (½ oz) butter or margarine	2.5 ml (½ level tsp) granulated sugar
450 g (1 lb) fresh peas, shelled	salt and pepper

1 SHRED THE LETTUCE, removing any thick stalks, and set aside. Put the remaining ingredients and 15 ml (1 tbsp) water in a medium ovenproof serving bowl. Cover and cook on HIGH for 5–7 minutes or until the peas are cooked.

2 ADD THE SHREDDED lettuce and cook on HIGH for 30 seconds–1 minute or until warmed through. Serve immediately.

BABY CARROTS WITH WATERCRESS AND ORANGE
————————SERVES 4 as an accompaniment————————
▲

Whole carrots cook beautifully in the microwave but choose small, even sized ones for even cooking.

bunch of watercress	60 ml (4 tbsp) orange juice
450 g (1 lb) whole new carrots, scrubbed	salt and pepper
15 g (½ oz) butter or margarine	

1 WASH THE WATERCRESS and reserve a few sprigs to garnish. Cut away any coarse stalks. Chop the leaves and remaining stalks.

2 PUT THE WATERCRESS and carrots in a shallow dish. Dot the butter or margarine over the vegetables and spoon over the orange juice. Season with pepper only.

3 COVER AND COOK on HIGH for 10–12 minutes or until tender. Adjust the seasoning before serving.

Summer Puddings (page 67)

HOT SHREDDED CELERIAC AND CARROT

———SERVES 4 as an accompaniment———

▲

*Celeriac combines particularly well with carrot to provide this
unusual, yet simple to prepare, vegetable accompaniment.*

450 g (1 lb) celeriac, peeled
30 ml (2 tbsp) lemon juice
2 large carrots, scrubbed and trimmed

salt and pepper
30 ml (2 tbsp) snipped fresh chives

1 COARSELY GRATE THE celeriac into a large bowl. Add the lemon juice and
30 ml (2 tbsp) water and toss together to prevent discoloration. Coarsely
grate the carrots and mix with the celeriac.

2 COVER AND COOK ON HIGH for 10–12 minutes until tender, stirring
occasionally.

3 SEASON TO TASTE with salt and pepper and serve sprinkled with the
snipped chives.

BLACKCURRANT JELLY WITH FRESH FRUIT

———SERVES 4———

▲

*Full of flavour, blackcurrants provide the attractive colour in this
jelly. It has a refreshing sharp tang but if you prefer, add a little sugar
to sweeten.*

225 g (8 oz) blackcurrants, stringed
finely grated rind and juice of $\frac{1}{2}$ lemon
15 ml (1 level tbsp) gelatine
300 ml ($\frac{1}{2}$ pint) unsweetened apple juice

prepared fresh fruit in season, such as
    strawberries, kiwi fruit, oranges,
    raspberries, to serve
few mint sprigs, to decorate

1 PUT THE BLACKCURRANTS and lemon rind and juice in a medium bowl.
Cook on HIGH for 5–6 minutes or until the blackcurrants are soft, stirring
occasionally.

2 PUT THE GELATINE and half of the apple juice in a small bowl and leave to
soak for 1 minute. Cook on HIGH for 30–50 seconds until the gelatine is
dissolved, stirring frequently. Stir into the blackcurrant mixture with the
remaining apple juice.

3 POUR THE JELLY into four 150 ml ($\frac{1}{4}$ pint) wetted moulds or ramekins and chill for 3–4 hours until set.

4 WHEN SET, TURN out on to individual plates and arrange the prepared fruit attractively around the jellies. Decorate with mint sprigs, if wished.

SUMMER PUDDINGS
————————SERVES 2————————
▲

This recipe makes two individual summer puddings—which look so pretty.

5–6 thin slices of day old white bread
150 g (5 oz) strawberries
150 g (5 oz) raspberries

45 ml (3 level tbsp) granulated sugar
strawberries and raspberries, to decorate

1 CUT THE CRUSTS off the bread and cut the bread slices into neat fingers. Reserve about a quarter and use the rest to line the base and sides of two 150 ml ($\frac{1}{4}$ pint) ramekin dishes, making sure that there are no spaces between the bread.

2 HULL THE STRAWBERRIES and put into a medium bowl with the raspberries.

3 SPRINKLE WITH THE sugar and add 45 ml (3 tbsp) water. Cover and cook on HIGH for 5–7 minutes or until the sugar dissolves, the juices begin to flow and the fruit softens.

4 RESERVE ABOUT 45 ML (3 tbsp) of the juice and pour the remaining fruit and juice into the lined ramekins. Cover with the reserved bread.

5 PLACE A SMALL saucer with a weight on it on top of each pudding and refrigerate overnight.

6 To SERVE, TURN out on to two serving plates and spoon over the reserved juice. Decorate with strawberries and raspberries.

POACHED APPLES AND PEARS
───────SERVES 4───────
▲

Fruit poached in the microwave retains its shape, texture and flavour well. Cooked in cider, these spicy apple and pear slices can be served hot or cold.

300 ml (½ pint) dry cider
1 cinnamon stick
2 cloves

2 large eating apples
2 large firm pears
3 large fresh dates (optional)

1 PUT THE CIDER, cinnamon and cloves in a large ovenproof serving bowl. Cook on HIGH for 3–5 minutes or until boiling.

2 CORE AND THINLY slice the apples and pears and stir into the hot cider. Cover and cook on HIGH for 4–5 minutes or until tender, stirring once.

3 MEANWHILE, STONE THE dates, if using, and cut into thin slices lengthways. Sprinkle the dates on top of the poached fruit. Serve hot or cold.

POACHED DRIED FRUITS
───────SERVES 2───────
▲

Dried fruits cook beautifully in the microwave and there is no need to pre-soak them before cooking. Packets of mixed fruit can now be bought in most supermarkets. If you are unable to find it, use dried fruits such as apricots, pears, prunes, peaches and apples.

150 g (5 oz) dried mixed fruit salad
300 ml (½ pint) fresh orange juice

strip of lemon rind
natural yogurt or single cream, to serve

1 PUT THE DRIED fruit in a medium ovenproof serving bowl. Pour over the orange juice and 100 ml (4 fl oz) water, then add the lemon rind. Mix well together.

2 COVER AND COOK on HIGH for 8–10 minutes, or until the fruits are almost tender, stirring occasionally.

3 LEAVE TO STAND, covered, for 5 minutes, then serve warm or chilled with yogurt or cream.

CHAPTER 4

ROASTING

The same principles of roasting in a conventional oven apply when roasting in a microwave. Large joints of meat cannot be cooked in a microwave because the microwaves cannot reach the centre of the joint without the outside being overcooked.

Smaller joints however can be cooked by this method. Kebabs, when cooked conventionally, are grilled or barbecued but when cooked in a microwave they are cooked by the roasting method and therefore are included in this chapter.

1   When roasting in the microwave, boned and rolled meat cooks more evenly than other joints because the shape and thickness are uniform.

2   When roasting meat, use a roasting rack to keep the juices from the underside of the meat and help the meat to brown.

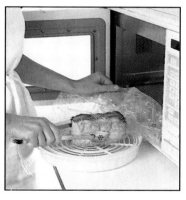

3   For even cooking, joints of meat should be uncovered halfway through, and then turned.

4   Kebabs should be repositioned during cooking to ensure even cooking.

5   Season meat at the end of cooking, never before, because salt draws out the moisture from the meat and makes it dry.

6   When roasting joints of meat, standing time must often be allowed. Remove the roast from the cooker, cover or wrap it in foil and leave to stand for the time specified in the recipe.

MINIATURE MONKFISH KEBABS ON A SPINACH PURÉE

SERVES 4 as a starter

▲

Monkfish has a firm, meaty texture that is perfect for kebabs.

450 g (1 lb) monkfish, skinned
15 ml (1 level tbsp) coriander seeds
5 ml (1 level tsp) ground cardamom
finely grated rind and juice of 1 lemon
15 ml (1 tbsp) vegetable oil

salt and pepper
450 g (1 lb) fresh spinach or 226 g (8 oz) packet frozen spinach
45 ml (3 tbsp) Greek strained yogurt
lemon wedges, to garnish

1 REMOVE AND DISCARD the central bone and cut the fish into twenty-four 2 cm ($\frac{3}{4}$ inch) cubes, the put in a shallow dish. Crush the coriander seeds and mix with the cardamom. Add the lemon rind and juice, the oil and salt and pepper to taste.

2 POUR THE SPICE mixture over the fish and mix so that all the fish is coated. Cover and leave to marinate for 20–30 minutes.

3 REMOVE ANY TOUGH stems from the fresh spinach, chop roughly and put in a bowl. Cover and cook on HIGH for 5–6 minutes until tender, stirring once. If using frozen spinach, put in a bowl, cover and cook on HIGH for 7–8 minutes, until thawed, stirring frequently. Drain thoroughly.

4 PURÉE THE SPINACH in a blender or food processor until smooth, then stir in the yogurt and season to taste with salt and pepper. Return to the bowl.

5 THREAD THE MONKFISH on to 12 wooden cocktail sticks, putting two pieces on each stick. Arrange in a single layer in the shallow dish containing the marinade. Cover and cook on HIGH for $2\frac{1}{2}$–$3\frac{1}{2}$ minutes until just tender, rearranging once.

6 POUR THE COOKING liquid from the fish into the spinach purée and cook on HIGH for 1–2 minutes or until hot. Spoon on to four plates. Arrange three kebabs on each plate on top of purée. Garnish and serve immediately.

CHICKEN IN MUSTARD AND LEMON SAUCE

SERVES 4 as a main course

▲

This tangy chicken dish uses a few simple ingredients, takes little time to cook, yet produces delicious results.

4 chicken breast fillets, skinned
20 ml (4 level tsp) wholegrain mustard
juice of 1 small lemon
15 ml (1 tbsp) vegetable oil

150 ml ($\frac{1}{4}$ pint) natural yogurt
salt and pepper
fresh herbs, to garnish

1 SLASH THE CHICKEN twice on one side, then spread all over with the mustard. Arrange in a single layer in a shallow dish. Sprinkle over the lemon juice and the oil.

2 ARRANGE IN A shallow dish and cook on HIGH for 7–8 minutes or until tender. Re-position the chicken once during cooking. Transfer the chicken to a warmed serving dish.

3 STIR THE YOGURT into the cooking dish and cook on HIGH for 1–1½ minutes until heated through, stirring once. Season to taste with salt and pepper. Pour over the chicken and serve immediately, garnished with fresh herbs.

CHICKEN SATAY
———————SERVES 4 as a main course or 6 as a starter———————
▲

Chicken Satay is a popular Indonesian dish consisting of tiny cubes of chicken served on wooden skewers with a spicy peanut sauce. Accompany with fried rice or noodles as a main course or with chunks of cucumber as a starter.

100 g (4 oz) creamed coconut	30 ml (2 tbsp) vegetable oil
90 ml (6 tbsp) crunchy peanut butter	2 garlic cloves, skinned and crushed
45 ml (3 tbsp) lemon juice	15 ml (1 tbsp) ground turmeric
30 ml (2 tbsp) soy sauce	5 ml (1 tsp) 5-spice powder
large pinch of chilli powder	5 ml (1 tsp) coriander seeds
4 chicken breast fillets, skinned	5 ml (1 tsp) cumin seeds

1 CRUMBLE 50 G (2 OZ) of the coconut into a medium bowl. Add the peanut butter, 15 ml (1 tbsp) of the lemon juice, 15 ml (1 tbsp) of the soy sauce, the chilli powder and 300 ml (½ pint) water. Cook on HIGH for 7–9 minutes until the sauce boils and thickens, stirring frequently. Turn into a small serving bowl.

2 CUT THE CHICKEN into small chunks and put in a bowl. Put the remaining coconut, lemon juice and soy sauce into a blender or food processor. Add the remaining ingredients and work until smooth.

3 POUR OVER THE chicken. Cover and marinate in the refrigerator for 2–3 hours or overnight.

4 THREAD THE CHICKEN on to 12 wooden skewers. Place in a shallow dish, pour over any remaining marinade and cook on HIGH for 10–12 minutes, turning frequently and basting with any remaining marinade. Serve hot, with the sauce for dipping.

CHICKEN TIKKA
——SERVES 4 as a main course——
▲

This Indian speciality, adapted here for the microwave, produces a wonderfully moist and spicy result. Accompanied by an onion, cucumber and pepper salad, it is equally good served hot or cold.

4 chicken breast fillets, skinned
juice of $\frac{1}{2}$ lemon
2.5 cm (1 inch) piece fresh root ginger, peeled and finely grated
1 green chilli, seeded and chopped
2 garlic cloves, skinned and crushed
5 ml (1 level tsp) garam masala
5 ml (1 level tsp) paprika
5 ml (1 level tsp) ground turmeric
5 ml (1 level tsp) ground cumin
5 ml (1 level tsp) ground coriander
30 ml (2 tbsp) chopped fresh coriander

150 ml ($\frac{1}{4}$ pint) natural yogurt
lemon wedges, to garnish
FOR THE SALAD
2 medium red or Spanish onions, skinned
$\frac{1}{2}$ cucumber
1 small green pepper, seeded
30 ml (2 tbsp) chopped fresh coriander
juice of $\frac{1}{2}$ lemon
5 ml (1 tsp) olive oil
salt and pepper

1 CUT THE CHICKEN into 2.5 cm (1 inch) cubes and place in a shallow dish. Put the lemon juice, ginger, chilli, garlic, garam masala, paprika, turmeric, cumin, ground coriander, the fresh coriander and yogurt in a blender or food processor and process until smooth.

2 POUR OVER THE chicken, cover and leave to marinate in the refrigerator for at least 3 hours or overnight.

3 TO MAKE THE salad, halve the onions and slice very thinly. Cut the cucumber into thin slices, then cut each slice crossways to make very thin strips. Cut the green pepper into very thin slices.

4 MIX THE ONION, cucumber and pepper together and sprinkle with the fresh coriander. Mix the lemon juice with the oil, pour over the salad and toss together until well mixed. Season with salt and pepper to taste, cover and leave to marinate in the refrigerator until ready to serve.

5 WHEN READY TO serve, thread the chicken on to eight wooden skewers and arrange in a single layer in a shallow dish. Cook on HIGH for 6–7 minutes or until tender, turning and basting with the marinade once during cooking. Garnish with lemon wedges, then serve with the salad.

Roast Chicken (page 74)

Roast Chicken and Duckling

▲

Because they cook so quickly and the method of cooking uses moist rather than dry heat, the skins of roast chicken and duckling will not be very brown, nor will they be crisp. To help them brown a little and to reduce splattering, cover with a split roasting bag. Commercially prepared browning agents are available, or paprika, honey, soy sauce or Worcestershire sauce may be brushed or sprinkled on to the skin before cooking. To achieve a brown and crisp skin, simply cook under a hot grill for a few minutes.

1 oven-ready roasting chicken or 1 oven-
   ready duckling
salt and pepper

few fresh herbs (optional)
stuffing (optional)

1 SEASON THE INSIDE of the chicken or duckling with salt and pepper. Place herbs inside the chicken and stuff the neck end if wished.

2 TRUSS THE CHICKEN or duckling into a neat compact shape using fine string. Weigh the bird and calculate the cooking time allowing 8–10 minutes per 450 g (1 lb) for chicken and 7–9 minutes per 450 g (1 lb) for duckling.

3 STAND THE BIRD on a roasting rack, breast side down, and stand the rack in a shallow dish to catch the juices. Cover with a split roasting bag and cook for half of the calculated time.

4 TURN OVER, RE-COVER and continue to cook for the remaining time.

5 COVER TIGHTLY WITH foil and leave to stand for 10–15 minutes before serving.

6 BROWN AND CRISP under a hot grill, if liked.

DUCKLING IN SWEET AND SOUR SAUCE

————————————SERVES 2 as a main course————————————
▲

Serve this with a colourful vegetable stir-fry. While the duck is cooking, slice carrots, peppers, courgettes into wafer thin slices and then into matchstick strips. Toss together in a large shallow dish with a little oil and soy sauce to flavour. When the duck is cooked, put the vegetables into the oven and cook on HIGH for 2–3 minutes, stirring frequently.

2 duckling breast fillets, each weighing about 200 g (7 oz)
1 orange
30 ml (2 tbsp) soy sauce
15 ml (1 level tbsp) dark soft brown sugar
15 ml (1 tbsp) clear honey

15 ml (1 tbsp) red wine vinegar
5 ml (1 tsp) sherry
5 ml (1 level tsp) cornflour
pinch of ground ginger
salt and pepper

1 PUT THE DUCKLING in a large shallow dish. Cut the orange in half, squeeze the juice from one half and pour over the duck. Cut the other half into slices and reserve for the garnish.

2 MIX THE REMAINING ingredients with 30 ml (2 tbsp) water and pour over the duck. Cover and leave to marinate for at least 30 minutes, turning once.

3 REMOVE THE DUCKLING from the marinade, leaving the marinade in the dish. Prick the duckling skin using a fork. Place the duckling, skin side up, on a roasting rack in a large shallow dish. Cover with a split roasting bag and cook on HIGH for 5 minutes or until the skin is just starting to brown. Remove from the cooker and leave to stand.

4 COOK THE RESERVED marinade on HIGH for 2–3 minutes or until boiling, then add the duckling portions, skin side down. Re-cover and cook on LOW for 8–10 minutes or until the duckling is tender.

5 TRANSFER THE DUCKLING to a serving dish and carve into thick slices. Blend the cornflour to a smooth paste with a little water and stir into the sauce. Cook on HIGH for 2 minutes until boiling and thickened, stirring occasionally. Spoon over the duckling and garnish with the reserved orange slices.

Below: Sage and Bacon-stuffed Pork (page 79)
Opposite: Honey Roast Gammon (page 78)

CROWN ROAST OF LAMB WITH MUSHROOM STUFFING

──────────SERVES 4 as a main course──────────

▲

Crown roast of lamb is made from two chined best end necks of lamb joined together. Get your butcher to prepare it for you or do it yourself. To do this, bend the joints around fat side inwards to form a crown and sew together using strong cotton or fine string. If you like your lamb cooked just until pink, cook for the lesser time—the longer time will give a well-done result.

25 g (1 oz) butter or margarine
1 medium onion, skinned and finely chopped
100 g (4 oz) button mushrooms, chopped
finely grated rind of 1 lemon
100 g (4 oz) fresh white breadcrumbs
15 ml (1 tbsp) chopped fresh parsley

15 ml (1 tbsp) chopped fresh thyme or 5 ml (1 level tsp) dried
salt and pepper
1 egg, lightly beaten
1 crown roast of lamb, prepared weight about 1.1 kg (2½ lb), made up of 12–14 chops

1 PUT THE BUTTER or margarine, onion and mushrooms in a medium bowl. Cover and cook on HIGH for 5–7 minutes until the onion has softened. Mix in the lemon rind, breadcrumbs, parsley, thyme and salt and pepper to taste. Add sufficient egg to bind the mixture.

2 FILL THE CENTRE of the crown with the stuffing, then weigh the joint and calculate the cooking time allowing 9–11 minutes per 450 g (1 lb).

3 PLACE THE CROWN on a roasting rack in a large shallow dish. Cook on MEDIUM for the calculated time. Turn round halfway through the cooking time.

4 WRAP THE CROWN tightly in foil and leave to stand for 10 minutes before serving.

HONEY ROAST GAMMON

──────────SERVES 6–8 as a main course──────────

▲

Gammon can vary in its degree of saltiness. Ask your butcher whether he recommends if it should be soaked or not before cooking.

1.4 kg (3 lb) gammon or collar
30 ml (2 tbsp) clear honey

30 ml (2 tbsp) orange marmalade
few drops of Tabasco sauce

1 WEIGH THE GAMMON and calculate the cooking time, allowing 7–8 minutes per 450 g (1 lb). Put the gammon in a roasting bag. Seal the end and prick the bag in several places. Stand on a roasting rack and place in a large shallow dish. Cook on HIGH for the calculated cooking time.

2 FIVE MINUTES BEFORE the end of the cooking time, remove the rind from the gammon and discard. Mix the honey, marmalade and Tabasco together and brush all over the joint. Continue cooking, uncovered, for the remaining time, brushing frequently with the marinade.

3 COVER TIGHTLY WITH foil and leave to stand for 10 minutes before serving hot or cold.

SAGE AND BACON-STUFFED PORK
SERVES 6–8 as a main course

▲

Gravy to serve with the joint is best made conventionally while the meat is standing. Transfer 30 ml (2 tbsp) fat (collected in the dish after cooking) to a saucepan, then stir in 20 ml (4 level tsp) flour. Cook for 2–3 minutes until brown, stirring all the time, then pour in 450 ml ($\frac{3}{4}$ pint) beef stock and season to taste with salt and pepper. Bring to the boil and simmer gently. Accompanying vegetables can be cooked in the microwave while the joint is standing.

about 1.8 kg (4 lb) loin of pork, boned and rinded
8 streaky bacon rashers, rinded
12 fresh sage leaves

2 garlic cloves, skinned and cut into slivers
salt and pepper
fresh sage, to garnish

1 PLACE THE PORK, fat side uppermost, on a flat surface and remove most of the fat. Score the remaining fat with a sharp knife.

2 TURN OVER THE meat and lay half of the bacon, the sage leaves and the garlic over the flesh. Season well with salt and pepper. Roll up and lay the remaining bacon on top.

3 SECURE THE JOINT with fine string. Weigh the joint and calculate the cooking time allowing 8 minutes per 450 g (1 lb). Place on a roasting rack, bacon side down, and cover with a split roasting bag. Stand the rack in a shallow dish to catch the juices. Cook on HIGH for half of the calculated cooking time, then turn over and cook for the remaining time.

4 WRAP TIGHTLY IN foil and leave to stand for 10 minutes. Serve cut into slices, garnished with fresh sage leaves.

KIDNEY AND BACON KEBABS
————————SERVES 4 as a main course————————
▲

Prick the kidneys to prevent them popping during cooking.

700 g (1½ lb) lamb's kidneys
8 streaky bacon rashers, rinded
100 g (4 oz) button mushrooms

vegetable oil
45 ml (3 tbsp) dry sherry
salt and pepper

1 REMOVE THE OUTER membranes from the kidneys and discard. Split each kidney in half lengthways and, using scissors, remove and discard the core. Prick each kidney twice with a fork.

2 STRETCH THE BACON rashers, using the back of a knife and cut each in half widthways. Roll up to make 16 bacon rolls.

3 THREAD THE KIDNEYS, bacon rolls and mushrooms on to eight wooden skewers. Arrange on a roasting rack in a single layer, then stand the rack in a large shallow dish. Brush with a little vegetable oil. Cook on HIGH for 8–9 minutes, rearranging and turning once.

4 REMOVE THE KEBABS from the cooker and transfer to a serving dish. Add the sherry to the juices collected in the dish and cook on HIGH for 3–4 minutes until boiling and slightly reduced. Season to taste with salt and pepper, then strain over the kebabs. Serve immediately.

HAKE AND LIME KEBABS
————————SERVES 4 as a light meal or a starter————————
▲

Kebabs are quick and simple to cook in a microwave but it is
important to use wooden kebab sticks. Always keep a supply at hand.

700 g (1½ lb) hake fillets, skinned
2 limes

salt and pepper

1 CUT THE HAKE into 2.5 cm (1 inch) cubes. Thinly slice one and a half limes. Thread the lime slices and the hake on to four wooden skewers. Arrange the kebabs in a single layer in a large shallow dish. Squeeze the juice from the remaining half lime over the kebabs.

2 COVER THE KEBABS and cook on HIGH for 4–5 minutes or until the fish is cooked, re-positioning the kebabs once during cooking. Season to taste with salt and pepper. Serve hot, accompanied with rice or cracked wheat.

Kidney and Bacon Kebabs

RED COOKED CHICKEN WITH
LETTUCE CHIFFONADE

————————SERVES 2 as a main course————————

▲

Chicken suprêmes are a French cut of skinned breast sold with the wing bone attached. You can find them in good butchers. If you have difficulty buying suprêmes, use skinned chicken breast fillets instead.

2 chicken suprêmes
5 ml (1 tsp) lemon juice
1 garlic clove, skinned and crushed
5 ml (1 level tsp) curry paste
15 ml (1 level tbsp) sweet paprika
15 ml (1 level tbsp) tomato purée

salt and pepper
5 ml (1 level tsp) cornflour
45 ml (3 tbsp) natural yogurt
½ small iceberg lettuce
½ small red pepper, seeded
15 ml (1 tbsp) olive oil

1 USING A SHARP knife slash the chicken at 1 cm (½ inch) intervals. Put in a small shallow dish. Sprinkle over the lemon juice and set aside.

2 To MAKE THE marinade, mix the garlic, curry paste, paprika, tomato purée, salt and pepper and cornflour together.

3 GRADUALLY BLEND IN the yogurt to make a thick paste. Spread the paste all over the chicken. Cover and leave to marinate in the refrigerator for at least 3 hours or overnight.

4 PLACE THE CHICKEN on a microwave roasting rack, arranging the thinner ends towards the centre. Cook on HIGH for 5 minutes or until tender.

5 WHILE THE CHICKEN is cooking, prepare the chiffonade. Shred the lettuce very finely and arrange on two serving plates. Season with salt and pepper. Slice the red pepper into very thin strips and arrange on top of the lettuce. Dribble over the olive oil.

6 SLICE THE CHICKEN and arrange on top of the chiffonade. Serve hot.

CHAPTER 5

QUICK FRYING

Quick frying covers both stir-fried and sautéed dishes, and these can be successfully adapted to the microwave cooker. As in conventional cooking, only *tender cuts of meat, poultry and vegetables are suitable. Quick frying uses only a small quantity of fat and the food is stirred frequently.*

1 Choose foods that cook quickly such as tender cuts of meat, chicken and vegetables.

2 Always use a large bowl or a large, round shallow dish for the foods to cook more evenly.

3 Cut the food to be cooked into small even-sized pieces for even cooking.

4 Add foods that cook more quickly last.

5 If the dish contains spices these should be fried for 1–2 minutes, as they are conventionally, to release their flavour.

6 Stir frequently, moving the food from the outside of the dish towards the centre to ensure even cooking.

COARSE HERB AND MUSHROOM PÂTÉ

———————————SERVES 6–8 as a starter———————————

▲

Use flat, black, open mushrooms for this pâté as they have the best flavour and give a good dark colour. If they are dirty, wipe them carefully with a damp cloth.

25 g (1 oz) butter or margarine
1 garlic clove, skinned and crushed
2 juniper berries, crushed
700 g (1½ lb) mushrooms, roughly chopped
75 g (3 oz) fresh brown breadcrumbs

60 ml (4 tbsp) chopped fresh mixed herbs
  such as thyme, sage, parsley, chervil
lemon juice
salt and pepper
fresh herbs, to garnish

1 PUT THE BUTTER or margarine, garlic and juniper berries in a large bowl and cook on HIGH for 1 minute.

2 ADD THE MUSHROOMS and cook on HIGH for 10–12 minutes or until the mushrooms are really soft and most of the liquid has evaporated, stirring frequently.

3 ADD THE BREADCRUMBS and herbs and season to taste with lemon juice and salt and pepper. Beat thoroughly together, then turn into a serving dish, cover and chill before serving. Garnish with fresh herbs and serve with Melba toast.

CHICKEN LIVER AND GREEN PEPPERCORN PÂTÉ

———————————SERVES 6 as a starter or light meal———————————

It's best to cover chicken livers during cooking or they will pop and splatter all over the cooker.

225 g (8 oz) chicken livers, finely chopped
100 g (4 oz) streaky bacon rashers, rinded
  and finely chopped
1 medium onion, skinned and finely
  chopped
15 ml (1 level tbsp) wholegrain mustard
15 ml (1 tbsp) brandy or sherry

1 garlic clove, skinned and crushed
10 ml (2 level tsp) green peppercorns,
  crushed
salt and pepper
100 g (4 oz) butter
lemon slices and parsley sprigs, to garnish

1 PUT THE LIVERS, bacon and onion in a large bowl with the mustard, brandy or sherry, garlic, green peppercorns and salt and pepper to taste.

2 COVER AND COOK on HIGH for 8 minutes, until the liver and bacon are tender, stirring frequently. Leave to cool.

3 PUT IN A blender or food processor with the butter and work until smooth. Adjust the seasoning.

4 SPOON INTO A serving dish, cover and chill in the refrigerator before serving. Garnish the pâté with lemon slices and parsley sprigs.

POTTED SHRIMPS
————————SERVES 4 as a starter————————
▲

Shrimps are very small crustaceans similar to small prawns. They are greyish brown in colour when alive and pink when cooked. They are not always available fresh, but can be bought frozen.

200 g (7 oz) butter
175 g (6 oz) cooked shrimps, peeled
pinch of ground mace
pinch of cayenne pepper

pinch of ground nutmeg
salt and pepper
bay leaves and peppercorns, to garnish

1 CUT HALF OF the butter into small pieces, put into a medium bowl and cook on HIGH for 1–2 minutes until melted.

2 ADD THE SHRIMPS, mace, cayenne pepper, nutmeg, salt and plenty of pepper. Stir to coat the shrimps in the butter, then cook on LOW for 2–3 minutes until the shrimps are hot, stirring occasionally. Do not allow the mixture to boil. Pour into four ramekin dishes or small pots.

3 CUT THE REMAINING butter into small pieces, put into a small bowl and cook on HIGH for 1–2 minutes until melted. Leave to stand for a few minutes to let the salt and sediment settle, then carefully spoon the clarified butter over the shrimps to cover completely. Garnish with bay leaves and peppercorns, leave until set, then chill in the refrigerator before serving.

4 SERVE STRAIGHT FROM the pots with brown bread and lemon wedges, or turn out and arrange on individual plates.

Below: Coarse Herb and Mushroom Pâté (page 84)
Opposite: Potted Shrimps (page 85)

COURGETTES TOSSED IN PARMESAN CHEESE

SERVES 4 as an accompaniment

▲

The courgette slices should be cooked until they are tender but still retain some bite.

450 g (1 lb) courgettes, trimmed
15 ml (1 tbsp) olive oil
1–2 garlic cloves, skinned and crushed

salt and pepper
25 g (1 oz) Parmesan cheese, freshly grated

1 CUT THE COURGETTES into 0.5 cm ($\frac{1}{4}$ inch) slices.

2 PUT THE OIL and garlic in a medium bowl and cook on HIGH for 2–3 minutes until the garlic is lightly browned, stirring occasionally.

3 ADD THE COURGETTES and toss to coat in the oil. Cook on HIGH for 4–6 minutes until the courgettes are just tender, stirring frequently.

4 SEASON TO TASTE with salt and pepper and sprinkle in the Parmesan cheese. Toss together until mixed, then serve hot.

CHERRY TOMATOES WITH PINE NUT AND BASIL DRESSING

SERVES 2 as an accompaniment

▲

Warm cherry tomatoes make an unusual hot accompaniment. If they are not available, use 225 g (8 oz) tomatoes, cut into quarters.

15 ml (1 tbsp) olive or vegetable oil
25 g (1 oz) pine nuts
2.5 ml ($\frac{1}{2}$ level tsp) Dijon mustard
2.5 ml ($\frac{1}{2}$ level tsp) light soft brown sugar

salt and pepper
2.5 ml ($\frac{1}{2}$ tsp) white wine vinegar
225 g (8 oz) cherry tomatoes, halved
15 ml (1 tbsp) chopped fresh basil

1 PUT THE OIL and pine nuts in a bowl and cook on HIGH for 2–3 minutes or until lightly browned, stirring frequently.

2 STIR IN THE mustard, sugar, salt and pepper to taste and whisk together with a fork. Whisk in the vinegar.

3 ADD THE TOMATOES and cook on HIGH for 30 seconds–1 minute, or until the tomatoes are just warm. Stir in the basil and serve immediately.

STIR-FRIED VEGETABLES
————————SERVES 4 as an accompaniment————————
▲

Stir frying is, by definition, a method of frying in shallow fat. The food must be cut into small, even-sized pieces and moved around constantly until coated. Stir-fried food is usually cooked in a wok and although this recipe isn't strictly stir-frying, it adapts well to the microwave. You can vary the vegetables according to your choice; for example, use broccoli florets, chopped celery or baby sweetcorn.

15 ml (1 tbsp) vegetable oil
15 ml (1 tbsp) soy sauce
30 ml (2 tbsp) dry sherry
1 garlic clove, skinned and finely chopped
2.5 cm (1 inch) piece of fresh root ginger, peeled and grated
2 medium carrots, sliced into matchstick strips

50 g (2 oz) beansprouts
100 g (4 oz) mange-tout, topped and tailed
1 red pepper, seeded and thinly sliced
4 spring onions, trimmed and chopped
½ head of Chinese leaves, thinly sliced

1 PUT THE OIL, soy sauce, sherry, garlic, ginger and carrots in a large bowl. Mix well together and cook on HIGH for 5 minutes or until the carrot is tender, stirring occasionally.

2 ADD THE REMAINING vegetables and mix together. Cook on HIGH for 5 minutes or until the vegetables are just tender, stirring frequently. Serve hot.

CHINESE CABBAGE WITH GINGER
————————SERVES 6 as an accompaniment————————
▲

Chinese cabbage is also known as Chinese leaves. It has tightly packed leaves with a sweet taste and crisp texture and it makes an excellent alternative to cabbage when cooked.

4 cm (1½ inch) piece fresh root ginger, peeled and thinly sliced
15 ml (1 tbsp) olive oil
15 ml (1 tbsp) soy sauce

large pinch of ground cloves
1 head of Chinese cabbage, trimmed and coarsely shredded
salt and pepper

1 PUT THE GINGER, oil, soy sauce and cloves in an ovenproof serving dish. Cook on HIGH for 2 minutes, stirring once.

2 ADD THE CHINESE cabbage and stir to coat in the oil. Season with salt and pepper to taste. Cook on HIGH for 3–4 minutes until hot but still crunchy, stirring occasionally. Serve immediately.

PIPERADE
SERVES 2 as a light meal

Covering the vegetables means that they cook quickly in a little fat and their own juices.

4 ripe tomatoes
25 g (1 oz) butter or margarine
1 small green pepper, seeded and chopped
1 garlic clove, skinned and crushed

1 small onion, skinned and finely chopped
salt and pepper
4 eggs, beaten
French bread, to serve

1 PRICK THE TOMATOES with a fork and cook on HIGH for 1½ minutes or until the skins burst. Peel off the skin, discard the seeds and roughly chop the flesh.

2 PUT THE BUTTER or margarine in a medium bowl and cook on HIGH for 45 seconds or until melted. Stir in the pepper, garlic and onion.

3 COVER AND COOK on HIGH for 3–4 minutes or until softened, stirring occasionally. Season to taste with salt and pepper and stir in the tomatoes. Cook on HIGH for 1 minute, then stir in the eggs.

4 COOK ON HIGH for 2–3 minutes or until the eggs are lightly scrambled, stirring frequently. Serve immediately with French bread.

SESAME CHICKEN WITH PEPPERS
SERVES 4 as a main course

Serve this Chinese stir-fry for a quick evening meal. Once the ingredients are prepared, it only takes a few minutes to cook. Chinese egg noodles or rice are the most suitable accompaniment.

4 chicken breast fillets, skinned
1 large red pepper
1 large yellow pepper
225 g (8 oz) can sliced bamboo shoots
6 spring onions, trimmed and sliced
2.5 cm (1 inch) piece of fresh root ginger, peeled and grated

30 ml (2 tbsp) vegetable oil
30 ml (2 tbsp) sesame seeds
30 ml (2 tbsp) soy sauce
30 ml (2 tbsp) dry sherry

1 CUT THE CHICKEN into thin strips. Cut the peppers into thin strips, discarding the core and seeds. Drain the bamboo shoots.

2 PUT ALL THE ingredients in a large bowl and stir well to mix. Cook on HIGH for 5–6 minutes until the chicken is tender and the vegetables are tender but firm, stirring occasionally. Serve hot.

Courgettes Tossed in Parmesan Cheese (page 88), Broad Beans with Bacon (page 28)

CHICKEN AND VEGETABLES WITH CASHEW NUTS

—————————SERVES 4 as a main course—————————

▲

Make sure that the chicken and vegetables are all cut to the same size to ensure even cooking.

75 g (3 oz) cashew nuts
3 chicken breast fillets, skinned
1 large green pepper, seeded
2 medium carrots
½ cucumber or 2 courgettes
½ head of Chinese leaves

30 ml (2 tbsp) soy sauce
1 garlic clove, skinned and crushed
1 cm (½ inch) piece fresh root ginger, peeled and grated
15 ml (1 tbsp) hoisin sauce

1 SPREAD OUT THE cashew nuts on a large flat ovenproof plate and cook on HIGH for 5 minutes until lightly browned, stirring frequently. Set aside.

2 MEANWHILE, CUT THE chicken, pepper, carrots, cucumber and Chinese leaves into thin shreds no more than 0.5 cm (¼ inch) wide.

3 PUT THE SHREDDED chicken and carrots in a large bowl with the soy sauce, garlic, ginger and hoisin sauce. Mix well together. Cook on HIGH for 5 minutes or until the chicken is tender, stirring occasionally.

4 ADD THE REMAINING ingredients and mix well together. Cook on HIGH for 5 minutes or until the vegetables are just tender, stirring frequently. Sprinkle with the cashew nuts and serve hot.

SHREDDED CHICKEN WITH MUSHROOMS AND POPPY SEEDS

—————————SERVES 2 as a main course—————————

▲

Poppy seeds give this dish an interesting speckled appearance. Serve with Stir-fried Vegetables (see page 89) and rice or noodles for a quick Chinese-style meal.

2 chicken breast fillets, skinned
15 ml (1 tbsp) vegetable oil
100 g (4 oz) button mushrooms, thinly sliced
7.5 cm (3 inch) piece of cucumber, cut into thin strips
4 spring onions, trimmed and chopped

25 ml (1 level tbsp) cornflour
30 ml (2 tbsp) sherry
150 ml (¼ pint) chicken stock
5 ml (1 tsp) white wine vinegar
10 ml (2 level tsp) black poppy seeds
salt and pepper
spring onions, to garnish

1 CUT THE CHICKEN into very thin strips. Put the oil in a large bowl and stir in the chicken. Cook on HIGH for $1\frac{1}{2}$–2 minutes or until the chicken changes colour, stirring occasionally.

2 STIR IN THE mushrooms, cucumber and onions and cook on HIGH for 3 minutes, stirring once.

3 MEANWHILE, BLEND THE cornflour with the sherry, stock and wine vinegar. Stir in the poppy seeds.

4 STIR THE LIQUID into the chicken and vegetables and cook on HIGH for 5–6 minutes or until thickened and the chicken is tender, stirring occasionally. Season to taste with salt and pepper and serve immediately, garnished with spring onions.

TURKEY STROGANOFF
SERVES 4 as a main course
▲

Do not allow the sauce to boil in step 3, or it will curdle. Serve this rich stroganoff with rice or pasta and a crisp green salad.

50 g (2 oz) butter or margarine
1 large onion, skinned and sliced
450 g (1 lb) turkey breast fillet, skinned and cut into thin strips
225 g (8 oz) mushrooms, sliced
150 ml ($\frac{1}{4}$ pint) white wine or chicken stock

150 ml ($\frac{1}{4}$ pint) soured cream
30 ml (2 level tbsp) tomato purée
15 ml (1 level tbsp) wholegrain mustard
10 ml (2 level tsp) paprika
1 egg yolk
salt and pepper

1 PUT THE BUTTER or margarine in a large bowl and cook on HIGH for 45 seconds or until the butter melts. Stir in the onion and cook on HIGH for 5–7 minutes until the onion is soft, stirring once.

2 STIR IN THE turkey, mushrooms and wine or stock. Cook on HIGH for 7 minutes or until the turkey is tender, stirring occasionally.

3 MIX THE REMAINING ingredients together, seasoning to taste with salt and pepper and add to the meat. Cook on MEDIUM for 4–5 minutes, until thickened, stirring after each minute. Do not allow to boil. Serve with rice or pasta.

Below: Beef with Ginger and Garlic (page 96)
Opposite: Pork with Fresh Plum Sauce (page 97)

BEEF WITH GINGER AND GARLIC

————————SERVES 2 as a main course————————

▲

*Cutting the carrots in wafer thin, ribbon-like strips means that they
cook quickly and look very pretty as well*

350 g (12 oz) fillet steak
2.5 cm (1 inch) piece of fresh root ginger,
  peeled and finely grated
1 garlic clove, skinned and crushed
150 ml ($\frac{1}{4}$ pint) dry sherry

30 ml (2 tbsp) soy sauce
2 medium carrots
15 ml (1 tbsp) vegetable oil
30 ml (2 level tbsp) cornflour
2.5 cm ($\frac{1}{2}$ level tsp) light soft brown sugar

1 CUT THE STEAK across the grain into 1 cm ($\frac{1}{2}$ inch) strips, and put into a
bowl. Mix the ginger with the garlic, sherry and soy sauce, then pour over
the steak, making sure that all the meat is coated. Cover and leave to
marinate for at least 1 hour.

2 USING A POTATO peeler, cut the carrots into thin slices lengthways.

3 PUT THE OIL in a large bowl and cook on HIGH for 1 minute or until hot.

4 USING A SLOTTED spoon, remove the steak from the marinade and stir
into the hot oil. Cook on HIGH for 1–2 minutes or until the steak is just
cooked, stirring once.

5 MEANWHILE, BLEND THE cornflour and the sugar with a little of the
marinade to make a smooth paste, then gradually blend in all of the
marinade.

6 STIR THE CARROTS into the steak and cook on HIGH for 1–2 minutes, then
gradually stir in the cornflour and marinade mixture. Cook on HIGH for 2–3
minutes until boiling and thickened, stirring frequently. Serve with rice.

MARINATED BEEF WITH MANGE-TOUT AND WALNUTS

————————SERVES 2 as a main course————————

▲

*Marinating meat in this dish not only improves the texture and
flavour but shortens the cooking time, too.*

175 g (6 oz) lean sirloin steak
30 ml (2 tbsp) dry sherry
30 ml (2 tbsp) soy sauce
1 garlic clove, skinned and crushed
1 cm ($\frac{1}{2}$ inch) piece fresh root ginger, peeled
  and grated

100 g (4 oz) mange-tout, topped and tailed
25 g (1 oz) walnuts, roughly chopped
pepper

1 TRIM THE MEAT of all excess fat, then cut across the grain into very thin strips about 5 cm (2 inches) long. Put in a medium bowl with the sherry, soy sauce, garlic and ginger. Cover and leave to marinate for at least 1 hour.

2 COOK ON HIGH for 3 minutes, stirring once.

3 ADD THE REMAINING ingredients, seasoning to taste with pepper and cook on HIGH for 3–4 minutes or until the beef is tender and the mange-tout just cooked, stirring occasionally. Serve hot.

PORK WITH FRESH PLUM SAUCE
————————SERVES 2 as a main course————————
▲

If preferred, the plum sauce can be made in advance and then reheated before the pork is cooked.

350 g (12 oz) pork fillet
50 ml (2 fl oz) chicken stock
50 ml (2 fl oz) fruity white wine
225 g (8 oz) fresh ripe red or purple plums,
  halved and stoned

15 ml (1 level tbsp) dark soft brown sugar
5 ml (1 tsp) lemon juice
salt and pepper
15 ml (1 tbsp) vegetable oil
fresh parsley sprigs, to garnish

1 CUT THE PORK into 1 cm ($\frac{1}{2}$ inch) slices. Place between sheets of greaseproof paper and flatten, using a meat mallet or a rolling pin, to a thickness of 0.5 cm ($\frac{1}{4}$ inch). Set aside.

2 TO MAKE THE SAUCE, put the stock and wine into a medium bowl and cook on HIGH for 5 minutes or until boiling and slightly reduced.

3 RESERVE TWO PLUM halves for the garnish, finely chop the remainder and stir into the hot liquid with the sugar and lemon juice. Cover and cook on HIGH for 3–4 minutes or until the plums are tender. Season to taste with salt and pepper.

4 ALLOW TO COOL a little, then purée the sauce in a blender or food processor until smooth. Pour back into the bowl and cook on HIGH for 5–7 minutes or until thickened and reduced.

5 PUT THE OIL in a shallow dish and cook on HIGH for 1–2 minutes or until hot. Stir in the pork and cook on HIGH for 4–5 minutes or until tender, turning once during cooking. Season to taste with salt and pepper.

6 REHEAT THE SAUCE on HIGH for 1–2 minutes or until hot, then spoon on to two warmed plates. Arrange the pork on the sauce, garnish with the reserved plum halves and the parsley sprigs and serve immediately.

LIVER WITH ORANGES
————————SERVES 4 as a main course————————

▲

Liver contains the mineral iron which is essential for good health. When eaten with food containing vitamin C, such as oranges, iron is more readily absorbed.

2 oranges
450 g (1 lb) lamb's or calf's liver, sliced
pepper

30 ml (2 tbsp) vegetable oil
1 medium onion, skinned and sliced
30 ml (2 tbsp) chopped fresh parsley

1 USING A SHARP knife, pare the rind from the oranges, then cut into thin strips. Alternatively, use a lemon zester.

2 PEEL THE ORANGES and divide into segments, discarding the pips.

3 CUT THE LIVER lengthways into pencil-thin strips, trimming away all ducts and gristle. Season to taste with pepper.

4 PUT THE OIL and onion in a shallow dish and cook on HIGH for 5–7 minutes until softened, stirring frequently.

5 STIR IN THE liver and half of the rind. Cook on HIGH for 3–4 minutes until the liver just changes colour, stirring occasionally.

6 STIR IN THE orange segments and parsley. Cook on HIGH for 3–4 minutes until the liver is tender and the oranges have softened slightly.

7 SERVE THE LIVER on a bed of hot green tagliatelle, garnished with the remaining rind strips.

Potato and Leek Ramekins (page 139), Cherry Tomatoes with Pine Nut and Basil Dressing (page 88)

STRIPS OF LIVER WITH FRESH HERBS

SERVES 4 as a main course

▲

Use lamb's liver for a family meal and calf's liver, considered the finest, for a special occasion. Although lamb's liver has a good flavour and texture, it is slightly coarser and darker than calf's.

15 ml (1 tbsp) vegetable oil
2 medium onions, skinned and sliced
450 g (1 lb) lamb's or calf's liver, sliced
100 g (4 oz) button mushrooms, sliced
30 ml (2 tbsp) chopped fresh mixed herbs,
  such as parsley, sage, tarragon

10 ml (2 level tsp) Dijon mustard
salt and pepper
fresh herb sprigs, to garnish

1 PUT THE OIL and the onions in a shallow dish and cook on HIGH for 7 minutes until softened, stirring occasionally.

2 MEANWHILE, CUT THE liver lengthways into pencil-thin strips, trimming away all ducts and gristle.

3 ADD THE LIVER to the dish and cook on HIGH for 3–4 minutes until the liver just changes colour, stirring occasionally.

4 STIR IN THE mushrooms, herbs and mustard, and season to taste with salt and pepper. Cover and cook on HIGH for 4–5 minutes until the liver is tender. Serve hot, garnished with sprigs of fresh herbs.

KIDNEYS AND MUSHROOMS IN SOURED CREAM SAUCE

SERVES 4 as a main course

▲

Do not overcook the kidneys or they will be tough. Stir frequently during cooking to make sure that they all cook at the same time, remembering that the kidneys nearest the edge of the dish will cook the quickest.

700 g (1½ lb) lamb's kidneys
30 ml (2 tbsp) vegetable oil
100 g (4 oz) button mushrooms
30 ml (2 tbsp) white wine

150 ml (¼ pint) soured cream
freshly grated nutmeg
salt and pepper

1 REMOVE THE THIN membrane from the kidneys and discard. Cut the kidneys in half lengthways and snip out the cores using kitchen scissors. Discard the cores.

2 PUT THE OIL in a large shallow dish and cook on HIGH for 1 minute or until hot. Add the kidneys and the mushrooms and toss together to coat in the oil. Cook on HIGH for 5–6 minutes or until cooked, stirring frequently.

3 TRANSFER THE KIDNEYS and mushrooms to a heatproof serving dish, then stir the wine into the juices left in the shallow dish. Cook on HIGH for 3 minutes or until boiling. Stir in the cream and cook on HIGH for 1 minute or until hot.

4 STRAIN THE SAUCE over the kidneys and mushrooms and season generously with nutmeg and salt and pepper. Cook on HIGH for 1 minute or until heated through, stirring occasionally. Serve immediately, with rice to mop up the juices.

SAUTÉED CHICKEN LIVERS
————————SERVES 3–4 as a main course————————
▲

Chicken livers must be pricked or, as in this recipe, chopped to prevent them from bursting and splattering when cooking them in a microwave.

15 ml (1 tbsp) vegetable oil
1 small onion, skinned and finely chopped
1 garlic clove, skinned and crushed
8 rashers of streaky bacon, rinded and
  chopped

450 g (1 lb) chicken livers
100 g (4 oz) mushrooms, sliced
60 ml (4 tbsp) natural yogurt
salt and pepper
chopped fresh parsley, to garnish

1 PUT THE OIL, onion, garlic and bacon in a large bowl. Cover and cook on HIGH for 5–7 minutes or until the onion is softened.

2 MEANWHILE, TRIM THE chicken livers and cut in half. Add the livers and mushrooms. Cook on HIGH for 5–7 minutes or until the livers are tender, stirring occasionally.

3 STIR IN THE yogurt and season to taste with salt and pepper. Cook on HIGH for 1–2 minutes or until hot. Serve garnished with chopped parsley. Accompany with boiled rice, if liked.

CHAPTER 6

PAN FRYING

It is not possible to shallow or deep fry in a microwave cooker, and these cooking techniques should never be attempted. However, by using a browning dish or skillet a similar result can often be achieved. Browning dishes and skillets are made of a material which can withstand a very high temperature. They are heated empty and the food to be cooked is then placed on the hot surface which immediately sears and browns it. Follow the manufacturer's instructions.

1   Heat the browning dish on HIGH for 5–8 minutes or according to manufacturer's instructions.

2   To ensure a crisp, brown surface use a maximum of 30 ml (2 tbsp) oil.

3   Do not remove the browning dish from the cooker as its temperature quickly reduces. Add the oil and food quickly to the dish as it sits inside the microwave.

4   Use oven gloves when removing the dish from the cooker as it becomes very hot.

5   Use tongs when repositioning food so that the fat is less likely to splatter.

6   When turning over food, place it on a different part of the dish so that the maximum heat from the dish is used.

ROSEMARY WALNUTS
———————SERVES 4–6 as a snack———————
▲

A tasty and satisfying alternative to bought peanut and crisp snacks. Use other fresh herbs if you prefer. Store in an airtight container and nibble as a snack or serve with drinks.

45 ml (3 tbsp) vegetable oil
15 ml (1 level tbsp) sweet paprika
350 g (12 oz) walnut halves

30 ml (2 tbsp) chopped fresh rosemary or
  15 ml (1 level tbsp) dried
salt (optional)

1 PUT THE OIL and paprika in a browning dish and cook on HIGH for $1-1\frac{1}{2}$ minutes until the oil is hot, stirring occasionally.

2 ADD THE WALNUTS and the rosemary and toss to coat in the oil. Cook on HIGH for 2–3 minutes until hot, stirring occasionally.

3 DRAIN ON A double sheet of absorbent kitchen paper to absorb any excess oil and sprinkle with salt, if liked. Serve warm or cold. Store in an airtight container for up to 2 weeks.

EGG AND BACON
——————— SERVES 1 ———————
▲

If you haven't got a hob, it's useful to know how to cook this in the microwave. Don't forget to prick the egg yolks or they will pop and splatter all over the cooker.

2 rashers back bacon, rinded
1 egg

1 HEAT A BROWNING dish on HIGH for 3–4 minutes only.

2 MEANWHILE, SNIP THE bacon fat at intervals. Quickly put the bacon rashers into the browning dish, keeping them towards the edge of the dish. Press down and cook on HIGH for 30–45 seconds.

3 TURN THE BACON over, then quickly break the egg into the centre of the dish. Prick the yolk using a cocktail stick or fine skewer and cook on HIGH for 45 seconds or until the egg is nearly set.

4 LEAVE TO STAND for 1 minute until the egg is set, then serve.

TINY CHEESE TRIANGLES
————————SERVES 4 as a starter————————
▲

Filo pastry, sometimes spelt phyllo, is a paper-thin pastry made of flour and water. It is possible to make it at home but it is a time-consuming task because of the amount of rolling and stretching needed to make it really thin, so buy it ready prepared and keep it in the freezer until required. When cooked in a browning dish these pastries are lightly browned and crisp.

75 g (3 oz) cream cheese
15 ml (1 tbsp) lemon or lime juice
1 spring onion, trimmed and finely chopped
25 g (1 oz) chopped dried apricots or dates
salt and pepper
75 g (3 oz) butter or margarine, cut into
  small pieces

4 sheets of packet filo pastry, thawed
75 ml (5 tbsp) natural yogurt
15 ml (1 tbsp) lemon juice
$\frac{1}{4}$ cucumber
mint sprigs, to garnish

1 To MAKE THE filling, mix the cream cheese and lemon or lime juice with the spring onion and chopped fruit and season to taste with salt and pepper.

2 PUT THE BUTTER or margarine in a small bowl and cook on HIGH for 2 minutes or until melted.

3 LAY ONE SHEET of pastry on top of a second sheet and cut widthways into six double layer 7.5 cm (3 inch) strips. Repeat with the remaining two sheets of pastry.

4 BRUSH THE STRIPS of pastry with the melted butter or margarine. Place a generous teaspoonful of filling at one end of each strip. Fold the pastry diagonally across the filling to form a triangle. Continue folding, keeping the triangle shape, until you reach the end of the strip of pastry. Repeat with the remaining strips of pastry to make a total of 12 triangles.

5 HEAT A BROWNING dish on HIGH for 5–8 minutes or according to the manufacturer's instructions.

6 MEANWHILE, BRUSH BOTH sides of each triangle with the melted butter or margarine.

7 USING TONGS, QUICKLY add six triangles to the dish and cook on HIGH for 1–2 minutes until the underside of each triangle is golden brown and the top looks puffy. Turn over and cook on HIGH for 1–2 minutes until the second side is golden brown.

8 REHEAT THE BROWNING dish on HIGH for 2–3 minutes, then repeat with the remaining triangles.

9 WHILE THE FILO triangles are cooking, make the sauce. Put the yogurt and lemon juice in a bowl and mix together. Grate in the cucumber and season to taste with salt and pepper.

10 SERVE THE FILO triangles warm or cold, garnished with mint sprigs, with the sauce handed round separately.

DEVILLED HERRINGS IN OATMEAL
SERVES 2 as a breakfast or light meal

▲

Herrings in oatmeal are a traditional breakfast or supper dish in Scotland. Here they are devilled to give them extra flavour.

10 ml (2 level tsp) tomato purée
2.5 ml ($\frac{1}{2}$ level tsp) mild mustard
2.5 ml ($\frac{1}{2}$ level tsp) brown sugar
dash of Worcestershire sauce
pinch of cayenne pepper
salt and pepper

4 small herring fillets
60 ml (4 level tbsp) medium oatmeal
15 ml (1 tbsp) vegetable oil
15 g ($\frac{1}{2}$ oz) butter or margarine
lemon wedges and mustard and cress, to garnish

1 HEAT A BROWNING dish on HIGH for 5–8 minutes or according to the manufacturer's instructions.

2 MEANWHILE, MIX THE tomato purée, mustard, sugar, Worcestershire sauce and cayenne pepper together. Season to taste with salt and pepper. Spread the paste thinly on to both sides of each fillet, then coat in the oatmeal.

3 PUT THE OIL and butter or margarine in the browning dish and swirl it around to coat the base of the dish.

4 QUICKLY ADD THE fillets, skin side down, and cook on HIGH for 1$\frac{1}{2}$ minutes. Turn over and cook on HIGH for 1–2 minutes or until the fish is cooked. Serve garnished with lemon wedges and mustard and cress.

WARM SALAD OF SALMON AND SCALLOPS

SERVES 4 as a light meal

▲

Although both salmon and scallops are expensive, this salad makes a little of each go a long way.

225 g (8 oz) salmon steak or cutlet
8 large shelled scallops
selection of salad leaves such as curly
    endive, Webb's wonder lettuce, radicchio
    and watercress
2 day-old bridge rolls
45 ml (3 tbsp) olive oil

45 ml (3 tbsp) crème fraîche or soured
    cream
10 ml (2 level tsp) wholegrain mustard
15 ml (1 tbsp) lemon juice
salt and pepper
a few chopped fresh herbs such as parsley,
    chives, dill and tarragon

1 SKIN THE SALMON and remove the bone, if necessary. Cut across the grain into very thin strips. If necessary, remove and discard from each scallop the tough white 'muscle' which is found opposite the coral. Separate the corals from the scallops. Slice the scallops into three or four pieces vertically. Cut the corals in half if they are large.

2 HEAT A BROWNING dish on HIGH for 5–8 minutes or according to the manufacturer's instructions.

3 MEANWHILE, TEAR THE salad leaves into small pieces, if necessary, and arrange on four plates. Cut the rolls into thin slices.

4 ADD 30 ML (2 TBSP) of the oil to the browning dish and swirl to coat the bottom of the dish. Quickly add the sliced rolls and cook on HIGH for 2 minutes. Turn over and cook on HIGH for a further 1 minute or until crisp. Remove from the dish and set aside.

5 ADD THE REMAINING oil and the scallops, corals and salmon to the dish and cook on HIGH for $1\frac{1}{2}$ minutes or until the fish looks opaque, stirring once.

6 USING A SLOTTED spoon, remove the fish from the dish, and arrange on top of the salad leaves.

7 PUT THE CRÈME fraîche or soured cream, mustard, lemon juice and salt and pepper to taste into the browning dish and cook on HIGH for 1–2 minutes or until hot. Stir thoroughly and pour over the fish. Sprinkle with the croûtons and herbs and serve immediately.

Tiny Cheese Triangles (page 104)

TROUT WITH ALMONDS
———————SERVES 2 as a main course———————
▲

Turn the fish over carefully, using tongs or it will break up and ruin the finished appearance of the fish.

2 rainbow trout, each weighing about 225 g (8 oz), cleaned
salt and pepper
15 ml (1 level tbsp) plain flour

15 ml (1 tbsp) vegetable oil
25 g (1 oz) butter or margarine, cut into pieces
25 g (1 oz) flaked almonds

1 HEAT A BROWNING dish on HIGH for 5–8 minutes or according to the manufacturer's instructions.

2 MEANWHILE, WIPE THE trout and cut off their heads just behind the gills. Wash and dry with absorbent kitchen paper, then season inside with salt and pepper. Season the flour with salt and pepper and use to coat the fish.

3 PUT THE OIL into the browning dish, then quickly add the fish. Cook on HIGH for 2 minutes, then turn over and cook on HIGH for 2 minutes or until the fish is cooked.

4 TRANSFER THE FISH to a serving dish and keep warm.

5 QUICKLY RINSE AND dry the browning dish, then add the butter or margarine and the almonds and cook on HIGH for 2–3 minutes or until lightly browned, stirring occasionally. Pour the almonds and butter over the trout and serve immediately.

FISHCAKES
———————MAKES 4———————
▲

Homemade fish cakes are far superior to the bought kind. This recipe uses freshly cooked fish, but is equally good made with ready cooked smoked fish such as trout or mackerel or canned fish such as tuna.

2 potatoes, each weighing 100 g (4 oz)
225 g (8 oz) fish fillets, such as smoked haddock, cod, salmon or coley
30 ml (2 tbsp) milk
25 g (1 oz) butter or margarine
finely grated rind of $\frac{1}{2}$ lemon
30 ml (2 tbsp) chopped fresh parsley

few drops anchovy essence (optional)
salt and pepper
beaten egg
30 ml (2 level tbsp) seasoned plain flour
30 ml (2 tbsp) vegetable oil
lime twists, to garnish

1 SCRUB THE POTATOES and prick all over with a fork. Cook on HIGH for 7 minutes, turning over once.

2 PUT THE FISH and the milk in a small shallow dish. Cover and put in the cooker with the potatoes. Cook on HIGH for 4–5 minutes until the fish flakes easily and the potatoes are soft.

3 FLAKE THE FISH, discarding the skin, and put in a bowl with the cooking liquid. Cut the potatoes in half, scoop out the flesh and add to the fish.

4 HEAT A BROWNING dish on HIGH for 5–8 minutes or according to the manufacturer's instructions.

5 MEANWHILE, MIX THE fish and potato with the butter or margarine, lemon rind, half the parsley, the anchovy essence, if using, and salt and pepper to taste. Mash thoroughly together, then mix with enough beaten egg to bind.

6 SHAPE INTO FOUR fishcakes about 2.5 cm (1 inch) thick. Mix the remaining parsley with the seasoned flour and use to coat the fishcakes.

7 ADD THE OIL to the browning dish, then quickly add the fishcakes and cook on HIGH for $2\frac{1}{2}$ minutes. Turn over and cook on HIGH for a further 2 minutes. Serve immediately, garnished with lime twists.

L A M B   C H O P S   W I T H   R O S E M A R Y   A N D   G A R L I C

————————————SERVES 2 as a main course————————————

▲

*Loin chops are recognisable by the small T bone. They are so tender
they cook well in a browning dish; serve still slightly pink in the centre.*

15 ml (1 tbsp) vegetable oil
4 lamb loin chops
25 g (1 oz) butter
1 small garlic clove, skinned and crushed
2.5 ml ($\frac{1}{2}$ tsp) finely chopped fresh rosemary
   or a large pinch of dried

15 ml (1 tbsp) lemon juice
salt and pepper
fresh rosemary sprigs, to garnish

1 HEAT A BROWNING dish on HIGH for 5–8 minutes or according to the manufacturer's instructions.

2 ADD THE OIL, then quickly add the chops. Cook on HIGH for $2\frac{1}{2}$ minutes, then turn over and cook on HIGH for $1\frac{1}{2}$ minutes or until cooked as desired.

3 TRANSFER THE CHOPS to a warmed serving dish. Stir the remaining ingredients into the browning dish, adding salt and pepper to taste, and cook on HIGH for $1\frac{1}{2}$ minutes until hot. Pour over the chops, garnish with rosemary sprigs and serve immediately.

PROVENÇAL LAMB FILLET

—————————SERVES 4 as a main course—————————

▲

Serve these lean slices of lamb fillet, cooked in a mouthwatering sauce of courgettes, garlic, tomato, herbs and red wine, with rice and a green salad.

2 medium courgettes
397 g (14 oz) can tomatoes
1 garlic clove, skinned and crushed
1 medium onion, skinned and finely
   chopped
15 ml (1 level tbsp) tomato purée
60 ml (4 tbsp) dry red wine

1 bay leaf
fresh thyme sprig or pinch of dried
few basil leaves or pinch of dried
salt and pepper
450 g (1 lb) lamb fillet
15 ml (1 tbsp) vegetable oil

1 CUT THE COURGETTES into 1 cm (½ inch) slices. Put in a large bowl with the tomatoes and their juice, garlic, onion, tomato purée, wine, herbs and salt and pepper to taste.

2 COOK ON HIGH for 15 minutes or until the sauce is slightly reduced and thickened, stirring once or twice during the cooking time.

3 MEANWHILE, CUT THE lamb into 1 cm (½ inch) slices. Cover with a piece of greaseproof paper and flatten them with a rolling pin.

4 HEAT A LARGE browning dish on HIGH for 5–8 minutes or according to the manufacturer's instructions.

5 WHEN THE BROWNING dish is ready, add the oil then quickly add the lamb. Cook on HIGH for 2 minutes or until lightly browned on one side. Turn over and cook on HIGH for 1–2 minutes or until the second side is brown.

6 POUR THE SAUCE over the lamb stirring to loosen any sediment at the bottom of the dish.

7 COOK ON HIGH for 3–4 minutes or until the lamb is tender, stirring occasionally. Serve hot.

Devilled Herrings in Oatmeal (page 105)

LAMB NOISETTES WITH ONION AND FRESH SAGE PURÉE

SERVES 2–4 as a main course

▲

Lamb noisettes are made from boned and rolled best end of neck. Look out for them in good butchers or supermarkets.

15 g ($\frac{1}{2}$ oz) butter or margarine
1 medium onion, skinned and finely
  chopped
75 ml (3 fl oz) chicken stock
2.5 ml ($\frac{1}{2}$ tsp) chopped fresh sage
5 ml (1 tsp) lemon juice
salt and pepper

45 ml (3 tbsp) soured cream
4 lamb noisettes, each about 4 cm
  (1$\frac{1}{2}$ inches) thick
15 ml (1 level tbsp) plain flour
15 ml (1 tbsp) vegetable oil
fresh sage leaves, to garnish

1 To MAKE THE purée, put the butter or margarine in a medium bowl and cook on HIGH for 30 seconds or until melted.

2 STIR IN THE onion, cover and cook on HIGH for 4–6 minutes or until really soft, stirring occasionally.

3 STIR IN THE stock, sage and lemon juice, re-cover and cook on HIGH for 3 minutes, stirring occasionally. Season to taste with salt and pepper. Leave to cool slightly, then add the soured cream.

4 HEAT A BROWNING dish on HIGH for 5–8 minutes or according to the manufacturer's instructions.

5 MEANWHILE, PURÉE THE onion mixture in a blender or food processor, then turn into a clean ovenproof serving bowl. Set aside.

6 LIGHTLY COAT THE noisettes with the flour and season with salt and pepper. Add the oil to the browning dish, then quickly add the noisettes, arranging them in a circle in the dish. Cook on HIGH for 2 minutes. Turn over and cook on HIGH for 1–2 minutes or until cooked as desired. They should still be slightly pink in the centre. Arrange on a warmed serving plate and garnish with fresh sage leaves.

7 COOK THE ONION purée on HIGH for 1–2 minutes or until hot and adjust the seasoning if necessary. Serve immediately with the noisettes.

Warm Salad of Salmon and Scallops (page 106)

ESCALOPES OF PORK IN MUSTARD CREAM SAUCE

SERVES 2 as a main course

▲

Some butchers sell ready prepared escalopes. If you cannot find them simply buy a 350 g (12 oz) piece of pork tenderloin, cut in half and flatten between sheets of greaseproof paper to make two thin escalopes.

15 ml (1 tbsp) vegetable oil
2 pork escalopes, each weighing about
   175 g (6 oz), rinded
1 small onion, skinned and finely chopped
100 g (4 oz) button mushrooms, sliced

75 ml (3 fl oz) dry white wine
15 ml (1 level tbsp) mild Dijon mustard
75 ml (3 fl oz) single cream
salt and pepper

1 HEAT A BROWNING dish on HIGH for 5–8 minutes or according to the manufacturer's instructions.

2 ADD THE OIL, then quickly add the pork and cook on HIGH for 2 minutes. Turn the escalopes over and cook on HIGH for 1 minute, then transfer to a warmed serving dish and keep warm.

3 STIR THE ONION and mushrooms into the browning dish and cook on HIGH for 3–4 minutes or until softened, stirring occasionally.

4 ADD THE WINE and cook on HIGH for 2 minutes or until slightly reduced, then stir in the mustard, cream and salt and pepper to taste and continue to cook on HIGH for 2 minutes or until reduced and thickened. Pour over the chops and serve immediately.

PORK WITH PINEAPPLE AND GREEN PEPPERCORNS

SERVES 2 as a main course

▲

The flavours of pork and pineapple complement each other perfectly, green peppercorns add an extra tang.

2 pork loin chops, each about 2.5 cm
   (1 inch) thick
226 g (8 oz) can pineapple slices in natural
   juice
5 ml (1 level tsp) cornflour

30 ml (2 tbsp) dry sherry
5–10 ml (1–2 level tsp) green peppercorns
salt
parsley sprigs, to garnish

1 HEAT A BROWNING dish on HIGH for 5–8 minutes or according to the manufacturer's instructions.

2 MEANWHILE, TRIM THE chops of excess fat and cut the fat into 2.5 cm (1 inch) pieces.

3 ADD THE FAT to the heated browning dish and cook on HIGH for 30 seconds until the fat starts to melt. Quickly add the chops, positioning the thinner ends towards the centre, and cook on HIGH for 2 minutes. Turn over and cook on HIGH for 1 minute. Remove the pieces of fat and discard.

4 DRAIN THE JUICES from the pineapple can into a bowl, then gradually blend in the cornflour. Stir into the browning dish with the sherry, green peppercorns and salt to taste. Cook on HIGH for 5 minutes or until the chops are tender, stirring occasionally.

5 ADD THE PINEAPPLE slices and cook on HIGH for 1–2 minutes or until heated through.

6 To SERVE, ARRANGE the chops on a warmed serving dish with the pineapple slices and spoon over the sauce. Garnish with parsley.

BEEFBURGERS
————————MAKES 4————————
▲

Traditionally, beefburgers contain nothing other than minced beef and seasoning, but they may be flavoured with a little grated onion or mustard if you prefer. Serve them rare or well done, according to your personal preference. Cook for the shorter time if you like rare burgers or the longer time if you prefer them well cooked.

450 g (1 lb) lean beef such as shoulder or rump steak, minced

salt and pepper
15 ml (1 tbsp) vegetable oil

1 HEAT A BROWNING dish on HIGH for 5–8 minutes or according to the manufacturer's instructions.

2 MEANWHILE, MIX THE beef with lots of salt and pepper. Shape into four burgers, 1.5 cm ($\frac{3}{4}$ inch) thick.

3 ADD THE OIL to the browning dish, then quickly add the burgers. Cook on HIGH for 3 minutes, then turn over and cook on HIGH for 2–4 minutes. Serve immediately.

Below: Fish Cakes (page 108)
Opposite: Lamb Noisettes with Onion and Fresh Sage Purée (page 112)

VEAL ESCALOPES WITH HAM AND MARSALA

————————————SERVES 2 as a main course————————————

▲

An Italian style veal dish. Serve with noodles to complete the theme.

1 veal escalope, weighing about 350 g
  (12 oz)
5 ml (1 tsp) lemon juice
salt and pepper
8 fresh sage leaves
4 thin slices prosciutto

25 g (1 oz) butter or margarine, cut into
  slices
15 ml (1 tbsp) vegetable oil
30 ml (2 tbsp) marsala
fresh sage leaves, to garnish

1 USING A ROLLING pin, flatten the escalope between two sheets of greaseproof paper. Cut into four.

2 HEAT A BROWNING dish on HIGH for 5–8 minutes or according to the manufacturer's instructions.

3 MEANWHILE, SPRINKLE THE escalopes with the lemon juice and season to taste with salt and pepper. Place two sage leaves on each escalope and cover each with a slice of prosciutto. Roll up and secure with a wooden cocktail stick.

4 PUT THE BUTTER and the oil in the browning dish, then quickly add the veal. Cook on HIGH for 2 minutes.

5 TURN OVER, RE-POSITION and cook on HIGH for 1 minute, then stir in the marsala and cook on HIGH for 2–3 minutes or until the meat is tender.

6 TRANSFER THE ESCALOPES to a warmed serving dish and remove the cocktail sticks. Cook the cooking juices on HIGH for 2–3 minutes or until reduced. Spoon over the escalopes and serve hot, garnished with fresh sage.

GRIDDLE SCONES

————————————MAKES 8————————————

▲

A microwave browning dish, skillet or griddle gives perfectly browned scones just like a conventional griddle.

225 g (8 oz) self raising flour
2.5 ml (½ level tsp) salt
15 g (½ oz) butter or margarine

25 g (1 oz) caster sugar
about 150 ml (¼ pint) milk or buttermilk
butter, for serving

1 HEAT A LARGE browning dish, skillet or griddle on HIGH for 4–5 minutes. Do not allow the dish to become too hot or the scones will burn.

2 PUT THE FLOUR and salt in a bowl. Rub in the butter or margarine, then stir in the sugar. Add enough milk or buttermilk to give a soft but manageable dough.

3 KNEAD LIGHTLY ON a floured surface, divide in two and roll into two rounds 0.5 cm ($\frac{1}{4}$ inch) thick. Cut each round into four.

4 QUICKLY PLACE FOUR quarters on the browning dish and cook on HIGH for 1$\frac{1}{2}$ minutes. Turn the scones over and cook on HIGH for a further 2 minutes. Repeat with the remaining scones, without reheating the browning dish. Eat while still hot, spread with butter.

VEGETARIAN BURGERS
MAKES 6
▲

Everyone likes these burgers, vegetarians and non-vegetarians alike. It is important to cook the spices to release their flavour.

2 medium potatoes, each weighing about 175 g (6 oz)
15 ml (1 level tbsp) coriander seeds
5 ml (1 level tsp) cumin seeds
30 ml (2 tbsp) vegetable oil
5 ml (1 level tsp) ground turmeric
1 garlic clove, skinned and crushed
100 g (4 oz) chopped mixed nuts
100 g (4 oz) Cheddar cheese, grated
1 egg yolk
30 ml (2 tbsp) chopped fresh coriander
salt and pepper

1 SCRUB THE POTATOES and prick all over with a fork. Cook on HIGH for 8 minutes or until tender, turning over once.

2 MEANWHILE, CRUSH THE coriander and cumin in a pestle and mortar.

3 WHEN THE POTATOES are cooked, remove from the cooker and set aside to cool slightly. Put half of the oil, the crushed spices, turmeric and garlic in a medium bowl and cook on HIGH for 2 minutes, stirring once.

4 PEEL THE SKINS from the potatoes and add to the spices, with the nuts, cheese, egg yolk, coriander and salt and pepper to taste. Mash thoroughly.

5 HEAT A BROWNING dish on HIGH for 5–8 minutes or according to the manufacturer's instructions.

6 MEANWHILE, USING LIGHTLY floured hands, shape the mixture into six burgers.

7 ADD THE REMAINING oil to the hot browning dish, then quickly add the burgers. Cook on HIGH for 2 minutes, then turn over and cook on HIGH for 2 minutes. Serve hot.

CHAPTER 7

SAVOURY BAKING

Baking in a conventional oven is by dry heat at the recommended temperature. Baking in a microwave cooker is by microwave energy at the recommended power setting. The food may be covered or uncovered depending on whether the end result is to have a moist or dry surface. Included in this chapter are the techniques for baking potatoes and other whole vegetables, baked eggs, fish and terrines.

1   Whole vegetables should be pricked to prevent them bursting.

2   Whole fish should be slashed to prevent them bursting.

3   Arrange several foods to be baked in a circle or square to ensure even cooking. Do not put anything in the centre as the microwaves penetrate the outside foods first.

4   Put whole foods, such as potatoes, on absorbent kitchen paper as this helps to absorb the moisture during cooking.

5   Thick foods such as potatoes should be repositioned and turned over during cooking because microwaves only penetrate the food to a depth of about 5 cm (2 inches).

6   Salt, if sprinkled directly on to foods such as meat, fish or vegetables, toughens and dries them out. It is therefore best to add salt after cooking.

EGGS FLORENTINE
————SERVES 4 as a light meal————
▲

It is very important to prick the egg yolks so that they do not burst, and even then to watch and listen for popping. If they do pop, quickly move their position in the cooker.

900 g (2 lb) fresh spinach, trimmed and coarsely chopped
25 g (1 oz) butter or margarine
45 ml (3 level tbsp) plain flour
1.25 ml ($\frac{1}{4}$ level tsp) mustard powder

300 ml ($\frac{1}{2}$ pint) milk
100 g (4 oz) Cheddar cheese, finely grated
salt and pepper
4 eggs
brown bread and butter, to serve

1 PUT THE SPINACH in a large bowl. Cover and cook on HIGH for 4 minutes or until the spinach is just tender. Leave to stand, covered.

2 PUT THE BUTTER or margarine, flour, mustard powder and milk in a medium bowl. Cook on HIGH for 4–5 minutes until the sauce has boiled and thickened, whisking after every minute. Stir in two thirds of the cheese. Season to taste with salt and pepper.

3 BREAK THE EGGS into four ramekin dishes or teacups. Gently prick the yolks with a fine skewer or the tip of a sharp knife and arrange in the cooker in a circle. Cook on HIGH for $1\frac{1}{2}$–2 minutes or until the egg whites are just set.

4 DRAIN THE SPINACH thoroughly, place in a flameproof dish, put the eggs on top and spoon the sauce over. Sprinkle with the reserved cheese and brown under a hot grill. Serve with brown bread and butter.

OEUFS EN COCOTTE

––––––––––SERVES 2 as a light meal––––––––––

▲

This recipe illustrates the technique of baking eggs in the microwave.

50 g (2 oz) button mushrooms
salt and pepper
4 drops of lemon juice
10 ml (2 tsp) chopped fresh parsley
2.5 ml ($\frac{1}{2}$ level tsp) plain flour

2 eggs
30 ml (2 tbsp) double cream
chopped fresh parsley, to garnish
buttered toast, to serve

1 ROUGHLY CHOP THE mushrooms and put into two 150 ml ($\frac{1}{4}$ pint) ramekin dishes. Cook on HIGH for 1 minute or until the mushrooms are almost cooked.

2 SEASON TO TASTE with salt and pepper and stir in the lemon juice and parsley. Sprinkle 1.25 ml ($\frac{1}{4}$ level tsp) flour into each ramekin and mix together well. Cook on HIGH for 30–45 seconds or until slightly thickened, stirring frequently.

3 BREAK THE EGGS into the dishes on top of the mushroom mixture. Gently prick the yolks with a cocktail stick or fine skewer.

4 COOK ON HIGH for 45 seconds–1 minute or until the whites are just set. Spoon over the cream and cook on HIGH for 30 seconds. Leave to stand for 2 minutes. Garnish with chopped parsley and serve hot with buttered toast.

AUBERGINE DIP WITH PITTA BREAD

––––––––––SERVES 2 as a starter or light meal––––––––––

▲

Aubergines baked in the microwave are simple to do and full of flavour. Don't forget to prick the skin or the aubergine will burst during cooking and make a mess in the cooker.

1 small aubergine
15 ml (1 tbsp) olive or vegetable oil
pinch of mild chilli powder
2.5 ml ($\frac{1}{2}$ level tsp) ground cumin
2.5 ml ($\frac{1}{2}$ level tsp) ground coriander
1 small garlic clove, skinned and crushed
10 ml (2 tsp) lemon juice

salt and pepper
150 ml ($\frac{1}{4}$ pint) natural yogurt
15 ml (1 tbsp) chopped fresh parsley
black olives and chopped fresh parsley, to garnish
2 pitta breads, to serve

1 WASH THE AUBERGINE and prick all over with a fork Cook on HIGH for 4–5 minutes or until very soft when pressed with a finger. Leave to stand.

2 MEANWHILE, PUT THE oil in a medium bowl, with the chilli powder, cumin, coriander and garlic. Cook on HIGH for 2 minutes, stirring occasionally. Stir in the lemon juice.

3 CUT THE AUBERGINE in half and scoop out the flesh. Mix with the cooked spices, mashing with a fork to make a pulp. Season well with salt and pepper and gradually beat in the yogurt. Stir in the chopped parsley.

4 SPOON THE AUBERGINE dip into two individual serving bowls and garnish with the black olives and parsley.

5 COOK THE PITTA bread on HIGH for 30 seconds or until warm. Cut into fingers and serve immediately with the aubergine dip.

PRAWN AND SESAME PARCELS
―――――――――――SERVES 2 as a snack―――――――――――
▲
These are rather like dimsum, the classic Chinese snack.

15 ml (1 tbsp) vegetable oil
50 g (2 oz) button mushrooms, chopped
1 cm (½ inch) piece of fresh root ginger, peeled and grated
2 spring onions, trimmed and finely chopped

75 g (3 oz) peeled prawns
10 ml (2 tsp) soy sauce
100 g (4 oz) strong wholemeal flour
15 ml (1 level tbsp) sesame seeds
salt and pepper
1 egg yolk

1 PUT THE OIL, mushrooms, ginger, spring onions, prawns and soy sauce in a medium bowl, and cook on HIGH for 2 minutes or until the mushrooms are softened.

2 PUT THE FLOUR, sesame seeds and salt and pepper to taste in a medium bowl. Make a well in the centre. Add the egg yolk and about 30 ml (2 tbsp) cold water to make a soft dough.

3 KNEAD THE DOUGH lightly, then roll out on a lightly floured surface to a 30.5 cm (12 inch) square. Cut into four squares, then divide the filling between them.

4 BRUSH THE EDGES of the pastry with water, then bring the four points of each square together and seal to form an envelope-shaped parcel.

5 PUT THE PARCELS on to a heatproof plate.

6 COVER WITH AN upturned bowl and cook on HIGH for 4–5 minutes or until the parcels are just set and firm to the touch. Serve immediately.

CHILLED COURGETTE MOUSSES WITH SAFFRON SAUCE

————————SERVES 2 as a starter————————

▲

Thin courgette slices make an attractive striped covering for this mousse. Be sure to cook the saffron sauce on LOW in step 6 or it will curdle.

275 g (10 oz) small courgettes, trimmed
15 g (½ oz) butter or margarine
7.5 ml (1½ tsp) lemon juice
100 g (4 oz) low fat soft cheese
salt and pepper

5 ml (1 level tsp) gelatine
45 ml (3 tbsp) natural yogurt
pinch of saffron strands
1 egg yolk
fresh herb sprigs, to garnish

1 USING A POTATO peeler or sharp knife, cut one of the courgettes into very thin slices lengthways. Put the slices into a medium bowl with 30 ml (2 tbsp) water. Cover and cook on HIGH for 2–3 minutes or until just tender, stirring once. Drain and dry with absorbent kitchen paper.

2 USE THE COURGETTE slices to line two oiled 150 ml (¼ pint) ramekin dishes. Set aside while making the filling.

3 FINELY CHOP THE remaining courgettes and put in a medium bowl with half of the butter or margarine and the lemon juice. Cover and cook on HIGH for 5–6 minutes or until tender, stirring occasionally.

4 ALLOW TO COOL slightly, then purée in a blender or food processor with the remaining butter or margarine and the cheese. Season well with salt and pepper.

5 PUT THE GELATINE and 15 ml (1 tbsp) water in a small bowl or cup and cook on LOW for 1–1½ minutes or until the gelatine has dissolved, stirring occasionally. Add to the courgette purée and mix together thoroughly. Pour into the lined dishes and leave to cool. Chill for at least 1 hour or until set.

6 MEANWHILE, MAKE THE sauce. Put the yogurt, saffron, egg yolk, salt and pepper in a small bowl and cook on LOW for 1–1½ minutes, or until slightly thickened, stirring frequently. Strain, then leave to cool.

7 TO SERVE, LOOSEN the courgette moulds with a palette knife, then turn out on to two individual serving plates. Pour over the sauce, garnish with a herb sprig and serve immediately.

Chilled Courgette Mousses with Saffron Sauce

BAKED POTATOES WITH CHILLI BEANS

SERVES 4 as a light meal

▲

Microwave baked potatoes are the perfect meal in a hurry, providing hot and filling high-fibre food in minutes. They are delicious topped simply with grated cheese, a dollop of natural yogurt and a generous seasoning of black pepper or fresh herbs such as chives, or coriander, or piled high with the filling suggested below.

397 g (14 oz) can chopped tomatoes
10 ml (2 level tsp) tomato purée
2 garlic cloves, skinned and crushed
2.5 ml ($\frac{1}{2}$ level tsp) chilli powder
2.5 ml ($\frac{1}{2}$ level tsp) dried oregano
425 g (14 oz) can red kidney beans, drained and rinsed

30 ml (2 tbsp) chopped fresh coriander or parsley
salt and pepper
4 large potatoes, each weighing about 175 g (6 oz)

1 PUT ALL THE ingredients, except the potatoes, in a large bowl and cook on HIGH for 10 minutes or until reduced and thickened.

2 SCRUB THE POTATOES and prick all over with a fork. Arrange on absorbent kitchen paper in a circle in the cooker and cook on HIGH for 12–14 minutes, or until the potatoes feel soft when gently squeezed, turning them over once during cooking.

3 REHEAT THE BEANS on HIGH for 2 minutes, stirring once.

4 CUT THE POTATOES in half and mash the flesh lightly with a fork. Pile the filling on top and serve immediately.

STUFFED PEPPERS

SERVES 4 as a light meal

▲

Although this recipe is for stuffed peppers, the filling could be used to stuff other vegetables such as baked potatoes or aubergines.

175 g (6 oz) burgul wheat
4 large green peppers
15 ml (1 tbsp) vegetable oil
2 garlic cloves, skinned and crushed
175 g (6 oz) Cheddar cheese, coarsely grated

1 large carrot, coarsely grated
1 large parsnip, coarsely grated
60 ml (4 tbsp) mayonnaise
chilli powder
salt and pepper

1 PUT THE WHEAT in a bowl and pour over 300 ml ($\frac{1}{2}$ pint) boiling water. Leave to soak for 10–15 minutes or until all the water is absorbed.

2 Cut the tops off the peppers and reserve. Scoop out the seeds and discard. Brush the peppers with the oil and stand upright in a dish just large enough to hold them.

3 Mix the remaining ingredients into the wheat and season to taste with chilli powder and salt and pepper. Use to stuff the peppers. Replace the reserved tops. Pour 30 ml (2 tbsp) water into the dish. Cover and cook on HIGH for 10–15 minutes or until the peppers are really tender. Serve hot or cold.

GREEK STUFFED AUBERGINES
SERVES 4 as a light meal

▲

These are delicious served warm, as they do in Greece, accompanied by a green salad.

30 ml (2 tbsp) olive oil
2 garlic cloves, skinned and crushed
5 ml (1 level tsp) ground allspice
5 ml (1 level tsp) ground cinnamon
30 ml (2 level tbsp) tomato purée
5 ml (1 tsp) mint sauce
50 g (2 oz) long-grain white rice

225 ml (8 fl oz) boiling vegetable stock
225 g (8 oz) minced lamb
2 medium aubergines, each weighing about 350 g (12 oz)
2 eggs, beaten
salt and pepper

1 Put 15 ml (1 tbsp) of the oil, the garlic, allspice and cinnamon in a large bowl and cook on HIGH for 1–2 minutes until the garlic softens.

2 Add the tomato purée, mint sauce, rice and stock, cover and cook on HIGH for 10–12 minutes or until the rice is tender and the liquid is absorbed.

3 Stir in the lamb and cook on HIGH for 5–10 minutes or until the meat changes colour, stirring occasionally.

4 Halve the aubergines lengthways, and scoop out the flesh, leaving a 1 cm ($\frac{1}{2}$ inch) shell. Finely chop the flesh and add to the meat mixture. Cook on HIGH for 5 minutes. Stir in the eggs and season to taste with salt and pepper.

5 Brush the aubergine halves, inside and out, with the remaining oil. Spoon in the filling. Arrange in an ovenproof serving dish and cook on HIGH for 8–10 minutes until the filling is set and the aubergine really soft. Serve warm with a salad.

VEGETABLE TERRINE
————————SERVES 6 as a light meal————————
▲

The finished terrine is marbled in appearance so looks very attractive.
You could use other vegetables, replace the potatoes with parsnips or
cauliflower, if liked.

700 g (1½ lb) carrots
700 g (1½ lb) potatoes
2 eggs
450 ml (¾ pint) double cream or Greek
strained yogurt
30 ml (2 tbsp) chopped fresh coriander or
parsley

salt and pepper
100 g (4 oz) Cheddar cheese, grated
25 g (1 oz) plain flour
25 g (1 oz) butter or margarine
300 ml (½ pint) milk
5 ml (1 level tsp) mild mustard
fresh coriander or parsley, to garnish

1 PEEL AND ROUGHLY chop the carrots and potatoes. Put into two roasting
bags with 15 ml (1 tbsp) water (keeping the two vegetables separate).
Loosely seal the bags and cook them, both at once, on HIGH for 15–17
minutes or until tender.

2 DRAIN THE CARROTS and put into a blender or food processor with 1 egg,
half of the cream or yogurt, the coriander or parsley and salt and pepper to
taste.

3 DRAIN THE POTATOES and mash with the remaining egg, cream or yogurt
and half of the cheese. Season to taste with salt and pepper.

4 SPOON ALTERNATE TABLESPOONFULS of the two mixtures into a greased
1.7 litre (3 pint) loaf dish. Level the surface.

5 STAND ON A roasting rack and cook on MEDIUM for 25 minutes or until
just firm to the touch. Leave to stand while making the sauce.

6 TO MAKE THE sauce, put the flour, butter or margarine, milk and
mustard in a bowl and cook on HIGH for 3–4 minutes until boiling and
thickened, whisking every minute. Stir in the remaining cheese and season
to taste with salt and pepper.

7 TURN OUT THE loaf on to a serving plate and garnish with coriander or
parsley. Serve sliced, with the sauce handed separately.

Lentil, Aubergine and Potato Pie (page 133)

STUFFED PLAICE TIMBALES WITH LEMON HERB BUTTER

————————SERVES 2 as a main course or 4 as a starter————————

▲

For this recipe you need two double plaice fillets, or one plaice, filleted. If your fishmonger takes quarter or single fillets from a plaice, you will need four.

25 g (1 oz) butter
5 ml (1 tsp) lemon juice
30 ml (2 tbsp) chopped fresh parsley
salt and pepper
175 g (6 oz) mushrooms

15 ml (1 tbsp) vegetable oil
75 g (3 oz) long grain white rice
300 ml ($\frac{1}{2}$ pint) hot chicken stock
2 large double plaice fillets, skinned
parsley sprigs, to garnish

1 To MAKE THE lemon herb butter, put the butter in a small bowl and beat until soft. Add the lemon juice, half of the parsley and season well with salt and pepper. Beat together well. Push to the side of the bowl to form a small pat and chill while making the timbales.

2 FINELY CHOP THE mushrooms and put in a medium bowl with the oil. Cover and cook on HIGH for 2–3 minutes or until the mushrooms are softened.

3 STIR IN THE rice and the chicken stock, re-cover and cook on HIGH for 10–12 minutes or until the rice is tender and the stock has been absorbed, stirring occasionally.

4 MEANWHILE, CUT THE plaice fillets in half lengthways, to make two long fillets from each. Place one fillet, skinned side in, around the inside of each of four buttered 150 ml ($\frac{1}{4}$ pint) ramekin or individual soufflé dishes. The fish should line the dish leaving a hole in the centre.

5 WHEN THE RICE is cooked, stir in the remaining parsley and salt and pepper to taste. Spoon this mixture into the centre of each ramekin, pressing down well. Cover loosely with absorbent kitchen paper and cook on HIGH for 2–3 minutes or until the fish is cooked.

6 LEAVE TO STAND for 2–3 minutes, then invert the ramekin dishes on to serving plates. With the dishes still in place pour off any excess liquid, then carefully remove the dishes.

7 GARNISH THE TIMBALES with parsley sprigs, then serve hot, with a knob of the lemon herb butter on top of each.

LASAGNE
———————————SERVES 4–6 as a main course———————————
▲

*Look out for the boxes of lasagne labelled 'no pre cooking required',
which are available at most large supermarkets. This will save you
time and energy, because it can be layered in the dish straight from the
packet, but always be sure to use plenty of sauce between the layers or
the finished dish will be dry.*

15 ml (1 tbsp) vegetable oil
1 large onion, skinned and chopped
2 garlic cloves, skinned and crushed
1 large green pepper, seeded and chopped
450 g (1 lb) lean minced beef
100 g (4 oz) mushrooms, sliced
397 g (14 oz) can tomatoes
30 ml (2 level tbsp) tomato purée
5 ml (1 level tsp) dried oregano

150 ml ($\frac{1}{4}$ pint) beef stock
salt and pepper
50 g (2 oz) butter or margarine
50 g (2 oz) plain flour
568 ml (1 pint) milk
50 g (2 oz) freshly grated Parmesan cheese
freshly grated nutmeg
175 g (6 oz) about 12 sheets, 'no need to pre
  cook' lasagne

1 PUT THE OIL, onion, garlic, pepper and beef in a large bowl and cook on
HIGH for 5 minutes or until the meat just changes colour, stirring
occasionally.

2 ADD THE MUSHROOMS, tomatoes with their juice, tomato purée and
oregano. Mix well together, cover and cook on HIGH for 10–15 minutes or
until the meat is cooked, stirring occasionally. Stir in the stock and season
to taste with salt and pepper.

3 PUT THE BUTTER or margarine, flour and milk in a large bowl and cook on
HIGH for 5–6 minutes until boiling and thickened, whisking frequently. Stir
in half of the Parmesan cheese and season to taste with salt, pepper and
nutmeg.

4 SPOON HALF OF the meat mixture into a large rectangular dish. Cover
with one third of the cheese sauce, then arrange half of the pasta on top.

5 REPEAT THE LAYERS, then spoon the remaining cheese sauce on top of the
pasta. Sprinkle with the remaining Parmesan cheese.

6 COOK ON HIGH for 30 minutes or until the pasta is cooked. Brown the top
under a hot grill, if liked. Serve, straight from the dish, with a mixed salad,
if liked.

LENTIL, AUBERGINE AND POTATO PIE

SERVES 4 as a main course

▲

This substantial, hearty dish is a vegetarian version of shepherd's pie. Don't throw away the potato skins in step 3, but cut them into neat strips and grill for 2–3 minutes on each side until crisp. Eat as a snack or serve with a dip.

3 medium potatoes, each weighing about 225 g (8 oz), scrubbed
100 g (4 oz) split red lentils
1 medium onion, skinned and finely chopped
1 bay leaf
5 ml (1 level tsp) dried thyme
15 ml (1 level tbsp) tomato purée

1 small aubergine, roughly chopped
450 ml (¾ pint) boiling vegetable stock
100 g (4 oz) French beans, trimmed and cut into 2.5 cm (1 inch) lengths
60 ml (4 tbsp) milk
salt and pepper
25 g (1 oz) Parmesan cheese, freshly grated

1 PRICK THE POTATOES all over with a fork and arrange in a circle on a sheet of absorbent kitchen paper. Cook on HIGH for 10–15 minutes or until soft, turning over halfway through cooking. Set aside to cool slightly.

2 WHILE THE POTATOES are cooling, put the lentils, onion, bay leaf, thyme, tomato purée, aubergine and vegetable stock in a large bowl and mix well together. Cover and cook on HIGH for 20–25 minutes or until the lentils and aubergine are tender and most of the liquid is absorbed. Add the beans and cook on HIGH for 2 minutes.

3 MEANWHILE, CUT THE potatoes in half and scoop out the flesh into a bowl. Mash with the milk and season to taste with salt and pepper.

4 SPOON THE LENTIL and aubergine mixture into a flameproof serving dish. Spoon over the mashed potato and sprinkle with the cheese. Cook on HIGH for 1–2 minutes or until heated through, then brown under a hot grill, if liked. Serve hot with a green vegetable or salad.

Brie and Watercress Tarts (page 134)

BRIE AND WATERCRESS TARTS
MAKES 4
▲

Serve these tarts as an attractive starter or with a side salad as a light lunch dish.

100 g (4 oz) plain wholemeal flour	275 g (10 oz) ripe Brie
salt	45 ml (3 tbsp) double cream
75 g (3 oz) butter or margarine	freshly grated nutmeg
2 bunches watercress	salt and pepper

1 To MAKE THE pastry, put the flour and salt to taste in a bowl. Add 50 g (2 oz) of the butter or margarine and rub in until the mixture resembles fine breadcrumbs. Add 30–60 ml (2–4 tbsp) water and mix together using a round-bladed knife. Knead lightly to give a firm, smooth dough.

2 ROLL OUT THE dough thinly. Invert four 10 cm (4 inch) shallow glass flan dishes and cover the bases and sides with the dough. Cover and chill while making the filling.

3 PUT THE REMAINING butter or margarine in a medium bowl and cook on HIGH for 1 minute or until melted. Trim and discard the tough stalks from the watercress. Reserve a few sprigs to garnish and stir the remainder into the butter. Cook on HIGH for 1–2 minutes until just wilted.

4 REMOVE THE RIND from the cheese and cut into small pieces. Stir into the watercress with the cream. Cook on HIGH for 1–2 minutes until melted. Season to taste with nutmeg, salt and pepper.

5 To COOK THE tarts, uncover and prick all over with a fork. Arrange pastry side uppermost in a circle in the cooker and cook on HIGH for 2–3 minutes or until firm to the touch.

6 LEAVE TO STAND for 5 minutes, then carefully loosen around the edge and invert on to a large serving plate. Fill with the filling and cook on HIGH for 2–3 minutes or until warmed through. Garnish with watercress.

BAKED RED MULLET EN PAPILLOTE
SERVES 2 as a main course
▲

This simple method of cooking fish in greaseproof paper parcels works just as well with any other small, whole fish.

2 red mullet, each weighing about 175 g (6 oz), cleaned and scaled	2 parsley sprigs
salt and pepper	2 bay leaves
$\frac{1}{2}$ small onion, skinned and thinly sliced	2 slices of lemon

1 SLASH THE FISH on each side using a sharp knife. Season the insides with salt and pepper to taste. Use the onion, parsley, bay leaves and lemon slices to stuff the fish.

2 CUT TWO 30.5 CM (12 inch) squares of greaseproof paper. Place a fish on each piece and fold it to make a neat parcel, twisting the ends together to seal. Place on a large flat plate.

3 COOK ON HIGH for 3–4 minutes or until the fish is tender. Serve the fish in their parcels.

MARBLED FISH RING
──────SERVES 6──────
▲

This simple method of putting alternate spoonfuls of the two fish mixtures, one tuna and the other white fish, into a ring mould produces a most attractive result.

200 g (7 oz) can tuna fish
15 ml (1 level tbsp) tomato purée
15 ml (1 tbsp) lemon juice
2 egg whites
300 ml ($\frac{1}{2}$ pint) natural yogurt
salt and pepper
700 g (1$\frac{1}{2}$ lb) white fish fillet, such as haddock, cod, whiting, skinned

100 g (4 oz) cream cheese
30 ml (2 tbsp) chopped fresh tarragon or 10 ml (2 level tsp) dried
15 ml (1 tbsp) chopped fresh dill or 5 ml (1 level tsp) dried
150 ml ($\frac{1}{4}$ pint) natural yogurt, to serve
fresh dill, to garnish

1 DRAIN THE TUNA and put into a blender or food processor with the tomato purée, lemon juice, one of the egg whites and 150 ml ($\frac{1}{4}$ pint) of the yogurt. Work until smooth. Season to taste with salt and pepper. Turn into a bowl and set aside.

2 ROUGHLY CHOP THE white fish fillet and put into the blender or food processor with the remaining yogurt, egg white, cream cheese, half of the tarragon and half of the dill. Work until smooth, then season with pepper.

3 PLACE ALTERNATE SPOONFULS of the fish mixtures into a 1.1 litre (2 pint) ring mould, then draw a knife through the two mixtures in a spiral to make a marbled effect. Level the surface.

4 COVER LOOSELY WITH absorbent kitchen paper, then cook on HIGH for 4–5 minutes or until the surface feels firm to the touch. Cool, then chill.

5 WHEN READY TO serve, mix together the natural yogurt and remaining tarragon and dill. Season to taste with salt and pepper.

6 TURN OUT THE ring and wipe with absorbent kitchen paper to remove any liquid. Cut into thick slices, garnish with dill and serve with the sauce.

POTATO, COURGETTE AND GARLIC BAKE
———————SERVES 4–6 as an accompaniment———————
▲

By grating the potatoes, they cook quickly in the microwave and by combining them with the grated courgettes, their speckled appearance looks very attractive.

25 g (1 oz) butter or margarine
1 garlic clove
2 eggs
150 ml ($\frac{1}{4}$ pint) soured cream

450 g (1 lb) potatoes, peeled and grated
350 g (12 oz) courgettes, grated
salt and pepper

1 GREASE AND LINE the base of a 1.1 litre (2 pint) ring mould.

2 PUT THE BUTTER or margarine and garlic in a small bowl and cook on HIGH for 30 seconds or until melted.

3 BEAT THE EGGS in a medium bowl, then stir in the soured cream. Add the grated potatoes and courgettes and mix well together. Stir in the melted butter. Season to taste with salt and pepper.

4 TURN THE MIXTURE into the prepared mould. Cover and cook on HIGH for 15 minutes or until set.

5 LEAVE TO STAND for 5 minutes, then turn out on to a warmed serving plate. Serve sliced.

CORN-ON-THE-COB WITH HERB VINAIGRETTE
———————SERVES 4 as an accompaniment———————
▲

Try to buy corn cobs with the green husk still attached, it makes a natural wrapping for baking and looks good left on for serving.

4 corn-on-the-cob
45 ml (3 tbsp) olive oil
30 ml (2 tbsp) lemon juice

30 ml (2 tbsp) chopped fresh mixed herbs
salt and pepper

1 PEEL BACK THE husks from the corn and remove the silk, then pull back the husks again to cover. If the corn is without husks, wrap separately in greaseproof paper.

2 PLACE THE CORN cobs side by side in a shallow dish. Cook on HIGH for 8–10 minutes until the corn is tender, turning and re-positioning two or three times during cooking.

3 MEANWHILE, WHISK THE oil, lemon juice and herbs together and season to taste with salt and pepper.

4 WHEN THE CORN is cooked, place on four warmed serving plates and gently pull back the husks or remove the greaseproof paper. Pour a little dressing over each cob and serve immediately.

POTATO AND LEEK RAMEKINS
SERVES 2 as an accompaniment

▲

Pretty to look at and delicious to eat, these ramekins are ideal to serve as a vegetable accompaniment when entertaining.

1 large potato, weighing about 225 g (8 oz)	freshly grated nutmeg
1 small leek	1 egg yolk
45 ml (3 tbsp) milk	15 g ($\frac{1}{2}$ oz) butter or margarine
salt and pepper	5 ml (1 level tsp) poppy seeds

1 GREASE AND LINE the bases of two 150 ml ($\frac{1}{4}$ pint) ramekin dishes with greaseproof paper.

2 PRICK THE POTATO all over with a fork, place on absorbent kitchen paper and cook on HIGH for 5–6 minutes or until soft, turning over halfway.

3 MEANWHILE, FINELY CHOP the white part of the leek and slice the green part into very thin 4 cm ($1\frac{1}{2}$ inch) long strips. Wash separately and drain.

4 PUT THE WHITE leek into a medium bowl with the milk, cover and cook on HIGH for 2–3 minutes or until very soft, stirring occasionally.

5 CUT THE POTATO in half, scoop out the flesh and stir into the cooked leek and milk. Mash well together and season to taste with salt, pepper and nutmeg. Stir in the egg yolk.

6 SPOON THE MIXTURE into the prepared ramekin dishes. Cover with a plate and cook on HIGH for 2–2$\frac{1}{2}$ minutes until firm to touch. Leave to stand.

7 MEANWHILE, PUT THE butter into a small bowl with the strips of green leek and the poppy seeds. Cover and cook on HIGH for 2–3 minutes or until tender, stirring occasionally. Season to taste with salt and pepper.

8 TURN THE RAMEKINS out on to a serving plate and spoon over the leek mixture. Cook on HIGH for 1–2 minutes to heat through. Serve hot.

CHAPTER 8

SAUCE MAKING

With a microwave cooker, sauces become quick and easy to make. White sauces, which are based on butter and flour, are less likely to become lumpy and even the more temperamental egg-based sauces such *as Hollandaise can be made with less danger of the sauce curdling. All can be made in one bowl and there is no risk of the sauce sticking to a pan or scorching.*

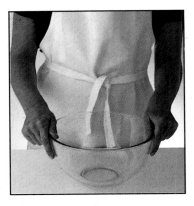 

1   Always use a container large enough to prevent the sauce from boiling over.

2   It is not necessary to cover the bowl when cooking a sauce because sauces need to be stirred frequently.

 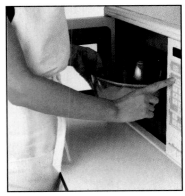 

3   Using a balloon whisk, whisk frequently to prevent lumps forming.

4   Most sauces can be cooked on a HIGH setting but sauces thickened with egg are best cooked on the LOW setting to prevent them curdling.

5   Sauces with a high sugar content attract microwaves and get very hot. Use oven gloves when removing the bowl from the cooker.

WHITE SAUCE
——MAKES 300 ml ($\frac{1}{2}$ pint)——
▲

A basic white sauce is wonderful cooked in a microwave. It is quick and simple to make because you use the all-in-one method. The secret for success is to whisk every minute, then you can guarantee a smooth sauce.

POURING SAUCE:
15 g ($\frac{1}{2}$ oz) butter or margarine
15 g ($\frac{1}{2}$ oz) plain flour
300 ml ($\frac{1}{2}$ pint) milk
salt and pepper

COATING SAUCE:
25 g (1 oz) butter or margarine
25 g (1 oz) plain flour
300 ml ($\frac{1}{2}$ pint) milk
salt and pepper

1 PUT ALL THE ingredients except the salt and pepper in a medium bowl and whisk together.

2 COOK ON HIGH for 4–5 minutes or until the sauce has boiled and thickened, whisking every minute. Season to taste with salt and pepper.

VARIATIONS
Add the following to the hot sauce with the seasoning:
Cheese sauce
50 g (2 oz) grated mature Cheddar cheese and a pinch of mustard powder.
Parsley sauce
30 ml (2 tbsp) chopped fresh parsley.
Hot tartare sauce
15 ml (1 tbsp) chopped fresh parsley, 10 ml (2 tsp) chopped gherkins, 10 ml (2 tsp) chopped capers and 15 ml (1 tbsp) lemon juice.
Caper sauce
15 ml (1 tbsp) capers and 5–10 ml (1–2 tsp) vinegar from the jar of capers.
Blue cheese sauce
50 g (2 oz) crumbled Stilton or other hard blue cheese and 10 ml (2 tsp) lemon juice.
Mushroom sauce
75 g (3 oz) sliced, lightly cooked mushrooms.
Onion sauce
1 medium chopped, cooked onion.
Egg sauce
1 finely chopped hard-boiled egg.

TOMATO SAUCE
——MAKES about 450 ml ($\frac{3}{4}$ pint)——
▲

Serve this simple tomato sauce with pasta. The sauce can be made in advance and reheated before serving.

30 ml (2 tbsp) olive oil
1 large onion, skinned and finely chopped
1 celery stick, trimmed and finely chopped
1 garlic clove, skinned and crushed
397 g (14 oz) can tomatoes

150 ml ($\frac{1}{4}$ pint) chicken or vegetable stock
15 ml (1 level tbsp) tomato purée
5 ml (1 level tsp) sugar
salt and pepper

1 PUT THE OIL, onion, celery and garlic in a large bowl. Cover and cook on HIGH for 5–7 minutes or until the vegetables are very soft.

2 STIR IN THE tomatoes with their juice, the stock, tomato purée and sugar. Season to taste with salt and pepper. Cook on HIGH for 10 minutes or until the sauce is reduced and thickened, stirring occasionally.

3 PURÉE IN A blender or food processor, pour back into the bowl and cook on HIGH for 2 minutes or until hot.

BREAD SAUCE
——MAKES 450 ml ($\frac{3}{4}$ pint)——
▲

Serve this traditional recipe hot with chicken, turkey and game dishes.

6 cloves
1 medium onion, skinned
4 black peppercorns
a few blades of mace
450 ml ($\frac{3}{4}$ pint) milk

25 g (1 oz) butter or margarine
100 g (4 oz) fresh breadcrumbs
salt and pepper
30 ml (2 tbsp) single cream (optional)

1 STICK THE CLOVES into the onion and place in a medium bowl together with the peppercorns and mace. Pour in the milk. Cook on HIGH for 5 minutes until the milk is hot, stirring occasionally.

2 REMOVE FROM THE cooker, cover and leave to infuse for at least 30 minutes.

3 DISCARD THE PEPPERCORNS and mace and add the butter or margarine and breadcrumbs. Mix well, cover and cook on HIGH for 3 minutes or until the sauce has thickened, whisking every minute. Remove the onion, season to taste with salt and pepper and stir in the cream, if using.

SAUCE HOLLANDAISE FOR POACHED SALMON
SERVES 4

▲

*Sauce Hollandaise is the classic sauce to accompany poached
salmon—but it can be difficult to make as it curdles very easily. Follow
these instructions exactly and you will achieve perfect results every
time. Another classic recipe using Hollandaise is Eggs Benedict. Cook
the eggs as shown in the recipe for Eggs Florentine on page 121. Put
the cooked eggs on toasted English muffins, topped with a slice of ham
and then spoon over the Hollandaise.*

4 salmon steaks, each weighing about 225 g (8 oz)
60 ml (4 tbsp) medium dry white wine
100 g (4 oz) butter, cut into small pieces

2 egg yolks
30 ml (2 tbsp) white wine vinegar
white pepper

1 ARRANGE THE SALMON with the thinner ends pointing towards the centre
in a large shallow dish. Pour over the wine, cover and cook on HIGH for
6–8 minutes or until tender. Leave to stand, covered, while making the
sauce.

2 To MAKE THE sauce, put the butter in a large glass bowl and cook on
HIGH for 30–60 seconds until just melted (do not cook for any longer or the
butter will be too hot and the mixture will curdle).

3 ADD THE EGG yolks and the vinegar and whisk together until well mixed.
Cook on HIGH for 1–1½ minutes, whisking every 15 seconds until thick
enough to coat the back of a spoon. Season with a little pepper.

4 TRANSFER THE SALMON to four serving plates and serve immediately,
with the sauce.

APPLE SAUCE
—— MAKES 150 ml ($\frac{1}{4}$ pint)——

▲

Serve this sauce with pork or sausages.

450 g (1 lb) cooking apples, peeled, cored
   and sliced
45 ml (3 tbsp) lemon juice

30 ml (2 level tbsp) caster sugar
25 g (1 oz) butter or margarine

1 PUT THE APPLES, lemon juice and caster sugar in a large bowl. Cover and cook on HIGH for 5–6 minutes or until the apples are soft, stirring frequently.

2 BEAT THE APPLES to a pulp with a wooden spoon or with a potato masher. If you prefer a smooth sauce, press the apples through a sieve or purée in a blender or food processor until smooth.

3 BEAT THE BUTTER or margarine into the apple sauce and spoon it into a serving bowl or jug. If the apples are very tart, add a little more sugar to sweeten to taste.

CRANBERRY SAUCE
——MAKES about 350 g (12 oz)——

▲

This is the traditional accompaniment to turkey but can also be served with other cold meats. The sugar should be added after the cranberries have cooked, otherwise the cranberries will be tough.

225 g (8 oz) fresh cranberries, stalks
   removed

225 g (8 oz) sugar
30 ml (2 tbsp) port (optional)

1 PUT THE CRANBERRIES in a large bowl and stab with the prongs of a fork so that most of the cranberry skins are pricked. This prevents the cranberries from bursting during cooking.

2 ADD 30 ML (2 TBSP) water, cover and cook on HIGH for 5 minutes, stirring frequently until thickened.

3 ADD THE SUGAR and mix well. Cook on HIGH for 2 minutes until the sugar is completely dissolved.

4 ADD THE PORT, if using. Allow to cool completely before serving.

GOOSEBERRY SAUCE
———————SERVES 2———————
▲

This sharp tangy gooseberry sauce makes the perfect accompaniment to mackerel. If you prefer a less sharp sauce, add a little more sugar.

175 g (6 oz) gooseberries, topped and tailed
15 ml (1 level tbsp) caster sugar
15 g ($\frac{1}{2}$ oz) butter or margarine

salt and pepper
freshly grated nutmeg

1 PUT THE GOOSEBERRIES, sugar, butter or margarine and 45 ml (3 tbsp) water in a medium bowl. Cover and cook on HIGH for 4–5 minutes until the gooseberries are softened, stirring once.

2 PURÉE IN A blender or food processor, then return to the rinsed-out bowl. Season well with pepper and nutmeg and add a little salt.

3 BEFORE SERVING, REHEAT the sauce on HIGH for 2 minutes or until hot.

BOLOGNESE SAUCE
–MAKES enough to dress 4 servings of pasta–
▲

This is the classic sauce to serve with pasta.

25 g (1 oz) butter or margarine
45 ml (3 tbsp) vegetable oil
2 rashers streaky bacon, rinded and finely chopped
1 small onion, skinned and finely chopped
1 small carrot, peeled and finely chopped
1 small celery stick, trimmed and finely chopped
1 garlic clove, skinned and crushed

1 bay leaf
15 ml (1 level tbsp) tomato purée
225 g (8 oz) lean minced beef
10 ml (2 tsp) chopped fresh herbs or 5 ml (1 level tsp) dried
150 ml ($\frac{1}{4}$ pint) dry red wine
150 ml ($\frac{1}{4}$ pint) beef stock
salt and pepper

1 PUT THE BUTTER or margarine and the oil in a large bowl and cook on HIGH for 1 minute. Stir in the bacon, vegetables and garlic and mix well. Cover and cook on HIGH for 6–8 minutes or until the vegetables begin to soften.

2 ADD THE BAY leaf to the vegetables and stir in the tomato purée and minced beef. Cook on HIGH for 3–4 minutes, stirring two or three times to break up the beef.

3 ADD THE HERBS, wine and stock and stir well to ensure that the meat is free of lumps. Cover and cook on HIGH for 4–5 minutes until boiling, then continue to cook on HIGH for 12–15 minutes or until the sauce is thick, stirring frequently. Season well with salt and pepper. Serve hot.

CURRY SAUCE
————SERVES 6————
▲

Serve this sauce with vegetables such as marrow or cabbage wedges, or mix with cooked fish, chicken or meat.

50 g (2 oz) butter or margarine, diced
1 medium onion, skinned and finely
   chopped
1 garlic clove, skinned and crushed
2.5 cm (1 inch) piece of fresh root ginger,
   peeled and finely chopped
10 ml (2 level tsp) ground turmeric
10 ml (2 level tsp) ground coriander

10 ml (2 level tsp) ground cumin
10 ml (2 level tsp) paprika
1.25 ml ($\frac{1}{4}$ level tsp) chilli powder
45 ml (3 level tbsp) plain flour
450 ml ($\frac{3}{4}$ pint) beef or chicken stock
30 ml (2 level tbsp) mango or apple chutney
salt and pepper

1 PUT THE BUTTER or margarine in a medium bowl and cook on HIGH for 1 minute until melted.

2 STIR IN THE onion, garlic and ginger and cook on HIGH for 5–7 minutes until softened.

3 STIR IN THE spices and flour and cook on HIGH for 30 seconds. Gradually stir in the stock.

4 COOK ON HIGH for 5–6 minutes until the sauce is boiling and thickened, whisking every minute.

5 ADD THE CHUTNEY and season to taste with salt and pepper. Cook on HIGH for 30 seconds to reheat.

SPINACH AND CHEESE SAUCE
————MAKES about 450 ml ($\frac{3}{4}$ pint)————
▲

This makes a delicate green spinach sauce. If liked, add a little Parmesan cheese for extra flavour. Serve with pasta of your choice.

15 ml (1 tbsp) vegetable oil
1 garlic clove, skinned and crushed
1 small onion, skinned and chopped
450 g (1 lb) fresh spinach, washed, trimmed
   and chopped, or a 226 g (8 oz) packet
   frozen spinach

100 g (4 oz) cream cheese
freshly grated nutmeg
salt and pepper

1 PUT THE OIL, garlic and onion in a medium bowl. Cover and cook on HIGH for 3–4 minutes or until softened.

2 STIR IN THE spinach. If using fresh, cover and cook on HIGH for 3–4 minutes or until the spinach is just cooked. If using frozen spinach, cook on HIGH for 8–9 minutes or until thawed. Drain.

3 PUT THE SPINACH in a blender or food processor and chop roughly. Add the cheese and purée until smooth. Season generously with nutmeg and salt and pepper.

4 RETURN TO THE bowl and cook on HIGH for 2–3 minutes or until hot. Serve with pasta of your choice.

BARBECUE SAUCE
————MAKES 300 ml ($\frac{1}{2}$ pint)————
▲

This sauce is ideal to serve with chicken, sausages, hamburgers or chops.

50 g (2 oz) butter or margarine
1 large onion, skinned and finely chopped
1 garlic clove, skinned and crushed
5 ml (1 level tsp) tomato purée
30 ml (2 tbsp) wine vinegar

30 ml (2 level tbsp) demerara sugar
10 ml (2 level tsp) mustard powder
1.25 ml ($\frac{1}{4}$ level tsp) chilli powder
30 ml (2 tbsp) Worcestershire sauce

1 PUT THE BUTTER or margarine in a medium bowl and cook on HIGH for 1 minute or until the butter melts.

2 STIR THE ONION and garlic into the melted butter, cover and cook on HIGH for 5–6 minutes until the onion softens.

3 WHISK ALL THE remaining ingredients together with 150 ml ($\frac{1}{4}$ pint) water and stir into the onion. Cook on HIGH for 5 minutes, stirring frequently. Serve hot.

147

SWEET WHISKED SAUCE
—————————MAKES 450 ml (¾ pint)—————————
▲

Serve this sauce spooned over fresh fruit or with fruit pies.

2 eggs, separated
25 g (1 oz) light soft brown sugar

30 ml (2 tbsp) white vermouth or sweet
  white wine

1 IN A MEDIUM bowl, using a hand-held electric mixer, whisk the egg yolks and sugar together until pale and creamy.

2 STIR IN THE vermouth or wine and cook on LOW for 2 minutes, whisking occasionally, until the mixture starts to thicken around the edges, then quickly remove from the cooker and whisk until smooth and thick.

3 WHISK THE EGG whites until stiff and fold into the sauce. Serve immediately.

CUSTARD
—————————MAKES 568 ml (1 pint)—————————
▲

*This familiar sauce is made in the same way as a savoury white sauce.
Serve it with puddings and pies.*

30 ml (2 level tbsp) custard powder or
  600 ml (1 pint) packet

15–30 ml (1–2 level tbsp) sugar
568 ml (1 pint) milk

1 BLEND THE CUSTARD powder and sugar with a little of the milk in a medium bowl. Stir in the remaining milk.

2 COOK ON HIGH for 3–4 minutes or until the sauce has thickened, stirring every minute. Stir well and serve hot or cold.

EGG CUSTARD SAUCE
————————MAKES 300 ml (½ pint)————————
▲

*Sauces thickened with egg are best cooked on a LOW setting as care is
needed to prevent them curdling. This is the classic custard sauce, made
with eggs. Serve with puddings and pies.*

300 ml (½ pint) milk
2 eggs

15 ml (1 level tbsp) granulated sugar
few drops of vanilla flavouring

1 POUR THE MILK into a large measuring jug and cook on HIGH for
2 minutes or until hot.

2 LIGHTLY WHISK THE eggs, sugar and vanilla flavouring together in a
bowl. Add the heated milk and mix well.

3 COOK ON HIGH for 1 minute, then cook on LOW for 4½ minutes or until the
custard thinly coats the back of a spoon. Whisk several times during
cooking. This sauce thickens slightly on cooling. Serve hot or cold.

JAM OR MARMALADE SAUCE
————————————SERVES 4————————————
▲

*Jam and marmalade have a high sugar content which attracts
microwaves so they get very hot. Watch the mixture carefully
therefore, and remove the bowl from the cooker using oven gloves.
Serve the sauce with steamed pudding.*

100 g (4 oz) jam or marmalade, sieved if
    preferred

2.5 ml (½ level tsp) cornflour
few drops of lemon juice

1 PUT THE JAM or marmalade and 150 ml (¼ pint) water in a medium bowl
and cook on HIGH for 2 minutes.

2 BLEND THE CORNFLOUR with 30 ml (2 tbsp) water, then stir into the
heated mixture.

3 COOK ON HIGH for 1–2 minutes until boiling, stirring after 1 minute. Add
lemon juice to taste, then serve hot.

CHOCOLATE SAUCE
————MAKES 300 ml (½ pint)————
▲

Who can resist hot chocolate sauce with ice cream?

15 ml (1 level tbsp) cornflour
15 ml (1 level tbsp) cocoa powder
30 ml (2 level tbsp) granulated sugar

300 ml (½ pint) milk
15 g (½ oz) butter

1 PUT THE CORNFLOUR, cocoa powder and sugar in a medium bowl and blend together with enough of the milk to give a smooth paste.

2 STIR IN THE remaining milk and the butter. Cook on HIGH for 3–4 minutes or until the sauce has thickened, stirring every minute. Stir well, then serve hot.

BUTTERSCOTCH SAUCE
————MAKES 150 ml (¼ pint)————
▲

When making a sauce with cornflour, make sure that it is completely dissolved in cold liquid before adding a hot one. Serve this sauce with ice cream and hot puddings.

170 g (6 oz) can evaporated milk
75 g (3 oz) soft brown sugar
25 g (1 oz) butter or margarine

2.5 ml (½ tsp) vanilla flavouring
15 ml (1 level tbsp) cornflour
25 g (1 oz) raisins (optional)

1 POUR THE EVAPORATED milk into a medium bowl and add 30 ml (2 tbsp) water and the brown sugar. Cook on HIGH for 3 minutes, stirring once. Add the butter or margarine and vanilla flavouring.

2 BLEND THE CORNFLOUR to a paste with a little cold water and add to the bowl, stirring well. Cook on HIGH for 2 minutes or until thickened, whisking once during the cooking time. Stir in the raisins, if using. Serve hot.

CHOCOLATE FUDGE SAUCE
————————MAKES 300 ml ($\frac{1}{2}$ pint)————————
▲

Golden syrup becomes very hot when cooked in a microwave so use oven gloves to remove the bowl from the cooker and watch that the mixture does not boil over. Serve with ice cream and other desserts.

75 ml (5 tbsp) single cream
25 g (1 oz) cocoa powder
100 g (4 oz) caster sugar
175 g (6 oz) golden syrup

25 g (1 oz) butter or margarine
pinch of salt
2.5 ml ($\frac{1}{2}$ tsp) vanilla flavouring

1 PUT ALL THE ingredients except the vanilla flavouring in a medium bowl and stir them together well.

2 COVER AND COOK on HIGH for 5 minutes or until boiling, stirring frequently.

3 STIR THE VANILLA flavouring into the sauce and allow it to cool slightly before serving.

HOT RASPBERRY SAUCE
————————MAKES 150 ml ($\frac{1}{4}$ pint)————————
▲

Serve hot with steamed puddings or try serving the sauce warm with ice cream.

225 g (8 oz) raspberries, sieved
45 ml (3 level tbsp) redcurrant jelly
15 ml (1 level tbsp) caster sugar

10 ml (2 level tsp) cornflour
5 ml (1 tsp) lemon juice

1 RUB THE RASPBERRIES through a nylon sieve into a medium bowl. Add the redcurrant jelly and caster sugar. Cook on HIGH for 2 minutes. Remove from the cooker and stir until the jelly has melted and the sugar has dissolved.

2 BLEND THE CORNFLOUR to a paste with 15 ml (1 tbsp) water and stir into the raspberry mixture. Cook on HIGH for 2 minutes or until thickened, whisking every 30 seconds. Stir in the lemon juice. Serve hot.

CHAPTER 9

PUDDING BAKING

Most puddings cook superbly in the microwave, especially steamed and suet puddings which become moist and light in a fraction of the time needed for conventional cooking. Fruits retain their colours and flavours when microwaved and they too cook in a fraction of the time taken to cook conventionally.

1   Whole fruits should be pierced to prevent them from bursting during cooking.

2   When cooking individual puddings, or fruits such as baked apples, arrange them in a circle with a space between each. Avoid placing one in the centre of the circle to ensure even cooking of the foods.

3   To test when a sponge pudding is cooked, it should be slightly moist on the top and a skewer inserted in the centre should come out clean.

4   Cook egg-based puddings on the LOW setting.

LAYERED FRUIT TERRINE
————————SERVES 6–8————————
▲

This looks impressive when sliced. Serve as a pretty dessert when entertaining.

100 g (4 oz) self raising flour
100 g (4 oz) softened butter or soft tub
  margarine
100 g (4 oz) light soft brown sugar
2 eggs
30 ml (2 tbsp) milk
275 g (10 oz) cream cheese
50 g (2 oz) caster sugar
50 g (2 oz) ground almonds
few drops of almond essence

300 ml ($\frac{1}{2}$ pint) double cream
15 ml (1 tbsp) gelatine
30 ml (2 tbsp) orange juice
3 kiwi fruits
225 g (8 oz) seedless white grapes, halved
225 g (8 oz) strawberries
15 ml (1 tbsp) icing sugar
15 ml (1 tbsp) orange-flavoured liqueur
  (optional)

1 GREASE A 1.7 LITRE (3 pint) loaf dish and line the base with greaseproof paper.

2 PUT THE FLOUR, butter or margarine, brown sugar, eggs and milk in a bowl and beat until smooth. Pour into the prepared loaf dish. Stand on a roasting rack and cook on HIGH for 4–5 minutes or until firm to the touch. Turn out and leave to cool on a wire rack.

3 MEANWHILE, BEAT THE cheese, caster sugar and ground almonds together. Flavour with almond essence. Whip the cream until it just holds its shape, then fold into the cheese mixture.

4 WHEN THE SPONGE is cold cut in half horizontally and return half to the bottom of the loaf dish.

5 PUT THE GELATINE and orange juice in a small bowl and cook on HIGH for 30 seconds–1 minute until dissolved; do not boil.

6 STIR INTO THE cheese mixture. Spread one third of the cheese mixture on top of the sponge lining the loaf dish. Peel and slice the kiwi fruits and arrange on top. Top with half of the remaining cheese mixture and then a layer of grapes. Cover the grapes with the remaining cheese mixture.

7 LEVEL THE SURFACE, then press the remaining piece of sponge on top. Chill in the refrigerator for 3–4 hours before serving.

8 TO MAKE THE sauce, purée the strawberries in a blender or food processor with the icing sugar and liqueur, if using. Serve the terrine sliced, with the strawberry sauce.

KIWI UPSIDE-DOWN PUDDING
———————SERVES 2———————

▲

You can use other fruit for this pretty pudding. Try sliced peaches or bananas for a change.

25 g (1 oz) butter or margarine
25 g (1 oz) light soft brown sugar
25 g (1 oz) self raising wholemeal flour
1.25 ml ($\frac{1}{4}$ level tsp) ground mixed spice

1 egg, beaten
2 kiwi fruits, peeled
15 ml (1 tbsp) clear honey
15 ml (1 tbsp) lemon juice

1 LINE THE BASE of a 7.5 × 11 cm (3 × $4\frac{1}{2}$ inch) ovenproof dish with greaseproof paper.

2 PUT THE BUTTER or margarine in a bowl and cook on HIGH for 10–15 seconds or until just soft enough to beat. Add the sugar, flour, mixed spice and the egg and beat well together, using a wooden spoon, until the mixture is well blended and slightly glossy.

3 CUT ONE OF the kiwi into thin slices and arrange in the base of the prepared dish.

4 CHOP THE REMAINING kiwi fruit and stir into the sponge mixture. Beat well together. Spoon the mixture on top of the kiwi slices and cover with a double thickness of absorbent kitchen paper.

5 COOK ON MEDIUM for 4–$4\frac{1}{2}$ minutes or until slightly shrunk away from the sides of the dish, but the surface still looks slightly moist. Leave to stand, covered, for 5 minutes, then turn out on to a serving plate.

6 MEANWHILE, PUT THE honey and lemon juice in a ramekin dish or cup. Cook on HIGH for 15–30 seconds or until warmed through. Spoon over the pudding and serve warm.

STUFFED BAKED APPLES
———————SERVES 4———————

▲

Nothing could be simpler than baking apples in a microwave but it is important to remember to cut the skins to prevent them from bursting.

4 medium cooking apples
clear honey
butter

1 CORE THE APPLES, then make a shallow cut through the skin around the middle of each.

2 STAND THE APPLES in a shallow ovenproof dish. Spoon a little honey into the centre of each apple and top with a knob of butter.

3 COOK ON HIGH for 5–7 minutes until the apples are tender. Turn the dish once during cooking. Leave to stand for 5 minutes, then serve with cream, yogurt or custard, if liked.

VARIATIONS
Omit the honey and stuff the apples with mincemeat or a mixture of dried fruits such as sultanas, currants and mixed peel or chopped dried dates, apricots or prunes, and flavour with grated orange, lemon or lime rind.

APPLE AND BLACKCURRANT CRUMBLE
SERVES 3–4

▲

This crumble has a rich, nutty texture and is far healthier than the conventional crumble as it contains little added sugar but gets most of its sweetness from coconut, nuts and sunflower seeds.

75 g (3 oz) butter or margarine
75 g (3 oz) plain wholemeal flour
25 g (1 oz) rolled oats
25 g (1 oz) sunflower seeds (optional)
15 g ($\frac{1}{2}$ oz) desiccated coconut
25 g (1 oz) chopped mixed nuts (optional)
25 g (1 oz) light soft brown sugar
5 ml (1 level tsp) ground cinnamon
  (optional)

2.5 ml ($\frac{1}{2}$ level tsp) ground mixed spice
  (optional)
225 g (8 oz) eating apples, cored and sliced
225 g (8 oz) blackcurrants
yogurt, cream or Custard, to serve (see
  page 148)

1 PUT THE BUTTER or margarine and flour into a bowl and rub together until the mixture resembles fine breadcrumbs. Stir in the dry ingredients and mix thoroughly together.

2 PUT THE APPLES and blackcurrants in a 1.1 litre (2 pint) deep ovenproof dish. Spoon the crumble mixture evenly over the fruit and press down lightly. Cook on HIGH for 11–12 minutes or until the fruit is tender. Serve hot or cold with yogurt, cream or custard.

ROLY-POLY PUDDINGS
―――――――――――――SERVES 4――――――――――――――
▲

The same basic suet pastry is used for jam roly-poly and all its variations. It is fast and easy to make with the help of a microwave cooker and, if mixed quickly and deftly, has a light, spongy texture—a far cry from the hefty steamed puds of schooldays.

175 g (6 oz) self raising flour
pinch of salt
75 g (3 oz) shredded suet

milk
Custard, to serve (see page 148)

1 MIX THE FLOUR, salt and suet together in a bowl.

2 USING A ROUND-BLADED knife, stir in enough water to give a light, elastic dough. Knead very lightly until smooth.

3 ROLL OUT TO an oblong about 23 × 28 cm (9 × 11 inches) and use as required. (See variations below). Brush the edges with milk and roll up, starting from the short end.

4 MAKE A 5 CM (2 inch) pleat across a large sheet of greaseproof paper. Wrap the roll loosely in the paper, allowing room for expansion. Pleat the open edges tightly together. Twist the ends to seal.

5 STAND THE PARCEL on a roasting rack and cook on HIGH for 4–5 minutes or until firm to the touch. Serve sliced, with custard.

VARIATIONS
Jam roly-poly
Spread the pastry with 60–90 ml (4–6 tbsp) jam.
Syrup roly-poly
Spread the pastry with 60 ml (4 tbsp) golden syrup mixed with 30–45 ml (2–3 tbsp) fresh white breadcrumbs.
Lemon roly-poly
Add the finely grated rind of 1 lemon to the dough. Roll out and spread with 60–90 ml (4–6 tbsp) lemon curd.
Mincemeat roly-poly
Add the finely grated rind of 1 orange to the dough. Roll out and spread with 60–90 ml (4–6 tbsp) mincemeat.
Spotted dick or dog
Replace half of the flour with 100 g (4 oz) fresh breadcrumbs. Add 50 g (2 oz) caster sugar, 175 g (6 oz) currants, finely grated rind of 1 lemon and 75 ml (5 tbsp) milk. Mix everything together. Shape into a neat roll about 15 cm (6 inches) long.

Apple and Blackcurrant Crumble (page 155)

CHRISTMAS PUDDING
——————SERVES 8——————
▲

A Christmas pudding can be cooked in a microwave but because the traditional Christmas pudding recipe contains a high proportion of sugar, dried fruits, fat and alcohol, all of which attract microwave energy and quickly reach a high temperature, it means great care must be taken not to overcook and possibly burn the pudding. As it is potentially dangerous, Christmas pudding should be watched during cooking. However, a Christmas pudding can be cooked in only 45 minutes in the microwave instead of 2½ hours by conventional cooking, and you do not need a saucepan of boiling water that has to be continually replenished. If, therefore, you are adapting your favourite Christmas pudding recipe, only add 30 ml (2 tbsp) of the alcohol suggested and replace the remaining liquid with milk or orange juice. Additional liquid should also be added to keep the pudding moist; allow an extra 15 ml (1 tbsp) milk for each egg added. A microwaved Christmas pudding should be eaten fairly soon after making. Store for up to 2–3 weeks in a cool place.

450 g (1 lb) mixed dried fruit	75 g (3 oz) fresh breadcrumbs
175 g (6 oz) stoned prunes	100 g (4 oz) shredded suet
450 ml (¾ pint) orange juice	100 g (4 oz) dark soft brown sugar
100 g (4 oz) plain flour	25 g (1 oz) blanched almonds, chopped
1.25 ml (¼ level tsp) freshly grated nutmeg	finely grated rind of ½ lemon
1.25 ml (¼ level tsp) ground cinnamon	30 ml (2 tbsp) sherry
2.5 ml (½ level tsp) salt	2 eggs, beaten

1 LINE THE BASE of a 1.3 litre (2½ pint) pudding basin with a circle of greaseproof paper.

2 PUT THE DRIED fruit, prunes and orange juice in a large bowl and mix well together. Cover and cook on HIGH for 20 minutes until the fruit is plump and the liquid absorbed. Leave to cool.

3 ADD THE REMAINING ingredients to the fruit mixture and mix well together. Spoon the mixture into the prepared basin, pushing down well.

4 COVER THE BASIN with a plate and cook on MEDIUM for 25–30 minutes until the top is only slightly moist.

5 LEAVE TO STAND, covered, for 5 minutes before turning out on to a warmed serving plate.

To reheat home-made and bought Christmas puddings

Christmas puddings containing a large quantity of alcohol or Christmas puddings that have previously been flambéed, are unsuitable for reheating in a microwave because of the risk of them catching fire (see above).

1 REMOVE ALL THE wrappings and basin from the pudding. Put the pudding on an ovenproof serving plate, cut into the required number of portions and pull apart so that there is a space in the centre.

2 PLACE A SMALL tumbler of water in the centre. This introduces steam and helps to keep the pudding moist. Cover with a large upturned bowl.

3 COOK ON HIGH for 2–3 minutes, depending on the size of the pudding, or until hot.

4 REMOVE THE COVER and glass and reshape the pudding with the hands. Decorate with a sprig of holly and serve.

5 To REHEAT AN individual portion of Christmas pudding, put on a plate and cook, uncovered, for 1–1½ minutes until hot.

ALMOND-STUFFED PEACHES
SERVES 4
▲

This is a light and refreshing pudding to serve in the summer. You can use nectarines instead of peaches if you prefer.

4 firm ripe peaches
50 g (2 oz) ground almonds
finely grated rind of ½ orange
5 ml (1 tsp) clear honey

150 ml (¼ pint) unsweetened orange juice
15 ml (1 tbsp) Amaretto (optional)
few mint leaves, to decorate

1 CUT THE PEACHES in half and carefully ease out the stones. Make the hollows in the peaches a little deeper with a teaspoon.

2 FINELY CHOP THE removed peach flesh and mix with the almonds, orange rind, honey and 15 ml (1 tbsp) of the orange juice.

3 USE THIS MIXTURE to stuff the peach halves, mounding the filling slightly.

4 PLACE THE PEACHES around the edge of a large shallow dish. Mix the remaining orange juice with the Amaretto, if using, and pour around the peaches.

5 COVER AND COOK ON HIGH for 3–5 minutes until the peaches are tender. Leave to stand, covered, for 5 minutes, then serve warm with the juices spooned over and decorated with mint leaves.

SPONGE PUDDING
SERVES 3–4

▲

Sponge puddings are wonderful cooked in a microwave; light and fluffy and quick to make too.

50 g (2 oz) soft tub margarine
50 g (2 oz) caster sugar
1 egg, beaten

few drops of vanilla flavouring
100 g (4 oz) self raising flour
45–60 ml (3–4 tbsp) milk

1 BEAT THE MARGARINE, sugar, egg, vanilla flavouring and flour until smooth. Gradually stir in enough milk to give a soft dropping consistency.

2 SPOON INTO A greased 600 ml (1 pint) pudding basin and level the surface.

3 COOK ON HIGH for 5–7 minutes until the top of the sponge is only slightly moist and a skewer inserted in the centre comes out clean.

4 LEAVE TO STAND for 5 minutes before turning out on to a warmed serving dish. Serve with custard.

VARIATIONS
Essex pudding
Spread jam over the sides and base of the greased pudding basin.
Apricot sponge pudding
Drain a 411 g (14½ oz) can of apricot halves and arrange them in the base of the greased pudding basin.
Syrup sponge pudding
Put 30 ml (2 tbsp) golden syrup into the bottom of the basin before adding the mixture. Flavour the mixture with the grated rind of a lemon.
Chocolate sponge pudding
Blend 60 ml (4 level tbsp) cocoa powder to a smooth cream with 15 ml (1 tbsp) hot water and add to the beaten ingredients.
Jamaica pudding
Add 50–100 g (2–4 oz) chopped stem ginger with the milk.
Lemon or orange sponge pudding
Add the grated rind of 1 orange or lemon when beating the ingredients.

FRESH FRUIT TARTLETS
————————MAKES 8————————

▲

*Take care when removing the pastry cases from the dishes as they are
very fragile.*

200 g (7 oz) plain flour
25 g (1 oz) plain wholemeal flour
75 g (3 oz) caster sugar
25 g (1 oz) ground toasted hazelnuts
pinch of salt
50 g (2 oz) butter or margarine
3 eggs, beaten
25 g (1 oz) cornflour

300 ml ($\frac{1}{2}$ pint) milk
few drops of vanilla flavouring
300 ml ($\frac{1}{2}$ pint) double cream
prepared fresh fruit, such as strawberries,
    raspberries, cherries, kiwi fruit, seedless
    grapes
30 ml (2 tbsp) apricot conserve
5 ml (1 tsp) lemon juice

1 PUT 175 G (6 oz) of the plain flour, the wholemeal flour, 25 g (1 oz) of
the sugar, the hazelnuts and salt into a bowl and mix together. Rub in the
butter or margarine until the mixture resembles fine breadcrumbs. Make a
well in the centre, add one egg and enough water to make a firm dough.

2 TURN ON TO a floured surface and knead for a few seconds until smooth.

3 CUT THE PASTRY in half, then roll out one half very thinly on a lightly
floured surface. Use to cover the base and sides of four inverted 10 cm
(4 inch) shallow glass flan dishes.

4 PRICK ALL OVER with a fork and cook on HIGH, pastry side uppermost, for
$2\frac{1}{2}$–3 minutes or until the pastry is firm to the touch. Remove the pastry
cases from the dishes and invert on to a wire rack to cool. Repeat with the
remaining pastry to make eight pastry cases.

5 To MAKE THE filling, put the remaining eggs and sugar in a large bowl
and whisk until pale and creamy and the mixture leaves a trail when the
whisk is lifted. Sift in the remaining flour and the cornflour, then beat well.

6 PUT THE MILK in a bowl and cook on HIGH for 2–$2\frac{1}{2}$ minutes until just
boiling. Gradually pour on to the egg mixture, stirring all the time. Add the
vanilla flavouring.

7 COOK ON HIGH for $1\frac{1}{2}$–2 minutes until very thick, stirring frequently.
Cover and leave to cool.

8 WHEN COLD, WHIP the cream until it just holds its shape, then fold into
the custard. Fill the pastry cases with the mixture and decorate with fruit.

9 PUT THE APRICOT conserve and lemon juice in a small bowl and cook on
HIGH for 30 seconds until melted. Brush over the tarts to glaze. Serve as
soon as possible.

Below: Layered Fruit Terrine (page 153)
Opposite: Baked Clementine Custards (page 166)

MICROWAVE CREAM MERINGUES
MAKES 32

▲

Conventional meringue mixture cannot be cooked in a microwave but by making a mixture similar to fondant the results are excellent. The mixture puffs up like magic and makes delicate meringues which can be topped with cream and fresh fruit.

1 egg white
about 275–300 g (10–11 oz) icing sugar
double cream, whipped

fresh fruit in season, such as strawberries, raspberries, kiwi fruit, peaches

1 PUT THE EGG white in a medium bowl and whisk lightly with a fork. Gradually sift in the icing sugar and mix to give a very firm, non-sticky but pliable dough.

2 ROLL THE MIXTURE into small balls about the size of a walnut. Place a sheet of greaseproof paper in the base of the cooker or on the turntable and arrange eight balls of paste in a circle on the paper, spacing well apart.

3 COOK ON HIGH for 1½ minutes until the paste has puffed up and formed meringue-like balls.

4 CAREFULLY LIFT THE cooked meringues off the paper and transfer to a wire rack to cool. Repeat three more times with the remaining fondant to make 32 meringues. Just before serving top with cream and fruit.

BAKED CLEMENTINE CUSTARDS
SERVES 2

▲

Baked custards can be successfully cooked in a microwave but it is important to cook them on a LOW setting and then leave to stand. During the standing time the custards continue to cook and set because of the heat retained in them.

2 clementines or satsumas
25 ml (1½ level tbsp) caster sugar
15 ml (1 tbsp) orange-flavoured liqueur
  (optional)

200 ml (7 fl oz) milk
1 egg and 1 egg yolk

1 FINELY SHRED THE rind of one of the clementines or satsumas. Put half into a heatproof jug with 10 ml (½ level tbsp) of the sugar and 75 ml (3 fl oz) water.

2 COOK ON HIGH for 2 minutes or until boiling, then continue to boil on HIGH for 2 minutes. Leave to cool.

3 PEEL AND SEGMENT the fruit, remove the pips and stir the fruit into the syrup with the liqueur, if using. Set aside to marinate.

4 MEANWHILE, MIX THE remaining rind and sugar with the milk, egg and egg yolk. Beat well together, then pour into two 150 ml ($\frac{1}{4}$ pint) ramekin or soufflé dishes. Cover, then cook on LOW for 8–10 minutes or until the custards are set around the edge but still soft in the centre.

5 LEAVE TO STAND for 20 minutes. When cool, chill for at least 2 hours. To serve, decorate with a few marinated clementine segments.

PEACH CHEESECAKE
————SERVES 12————
▲

Instead of a crunchy base, this cheesecake has a soft sponge base which is spooned on top of the cheese mixture before cooking. To prevent it sinking into the cheese, place teaspoonfuls of sponge mixture all over the top and then spread carefully with a knife.

3 ripe peaches, skinned and stones removed
450 g (1 lb) curd cheese
50 g (2 oz) caster sugar
15 ml (1 tbsp) lemon juice
2 eggs
15 ml (1 level tbsp) cornflour
300 ml ($\frac{1}{2}$ pint) soured cream

FOR THE BASE
50 g (2 oz) self raising flour
50 g (2 oz) softened butter or soft tub margarine
50 g (2 oz) light soft brown sugar
1 egg
15 ml (1 tbsp) milk

1 GREASE A DEEP 20.5 cm (8 inch) round dish and line the base with greaseproof paper. Grease the paper.

2 ROUGHLY CHOP THE peaches. Put half in a blender or food processor with the cheese, sugar, lemon juice, eggs, cornflour and half of the cream. Mix until smooth. Stir in the remaining chopped peaches, then pour into the prepared dish and level the surface.

3 PUT THE INGREDIENTS for the base in a bowl and beat together until smooth. Spoon carefully on top of the peach mixture, then level with a palette knife being careful not to disturb the cheese mixture.

4 STAND ON A roasting rack and cook on MEDIUM for 20 minutes until a skewer inserted in the centre comes out clean and the sponge mixture on top is risen. (The sponge will look very moist at this stage.)

5 LEAVE TO STAND for 15 minutes, then loosen around the sides with a palette knife and carefully turn out on to a flat serving plate so that the sponge base is at the bottom. Peel off the greaseproof paper. (If the cheesecake is still not quite cooked in the centre, return it to the cooker, on the serving plate and cook on MEDIUM for 1–2 minutes until set.)

6 SPREAD THE REMAINING soured cream on top. Leave until cool, then chill in the refrigerator for 2–3 hours or overnight before serving.

167

CHAPTER 10

CAKES, BISCUITS AND BREAD BAKING

Microwave cookers produce light, even-textured cakes but because the sugar in them does not caramelise and form a crust, and because the mixtures are always moist, the cakes do not brown. However, if the mixture contains ingredients that colour it, the lack of extra browning will not matter, *alternatively the cake can be decorated. The same applies when cooking bread, and to produce a crust, the loaf can be finished off under a grill. Traditional crisp biscuits cannot be produced in a microwave but good results can be obtained with moist cookie-type biscuits such as flapjacks.*

1   Cakes baked in a plastic container will not need greasing unless the mixture contains only a small amount of fat, but other containers should be greased and the base of larger containers lined with greaseproof paper. Avoid flouring dishes as this produces an unpalatable coating on the cake.

2   Large cakes are better if cooked in a ring mould otherwise the centre will not be cooked.

3   When cooking a number of small cakes in paper cases, use two per cake for extra support and arrange them in a circle about 5 cm (2 inches) apart for even cooking; do not put a cake in the centre. This ensures even penetration of the microwaves and therefore even cooking.

4   Cakes, that are not cooked in a ring mould, should be raised on a trivet during cooking so that the microwaves can penetrate the cake from all sides.

5   Even if the microwave has a turntable or stirrers, cakes that rise unevenly should be repositioned during cooking to ensure even cooking.

6   Remove cakes from the cooker when they are still moist on top (normally they would be considered slightly underdone) then leave for the time recommended in the recipe. During the standing time the cooking will be completed by the conduction of heat.

7   Use cocoa powder, chocolate, brown sugar, spices, brown flour or icing to overcome the fact that cakes and breads do not brown in the microwave.

SPICY APPLE CAKE
——————MAKES 16 slices——————
▲

This very moist cake is equally good served warm as a dessert with custard, cream or yogurt. It includes mixed spice and ground cinnamon to make a spicy cake, but if you prefer you can omit them and add a little finely grated lemon rind instead.

450 g (1 lb) cooking apples, peeled, cored and roughly chopped
225 g (8 oz) plain wholemeal flour
10 ml (2 level tsp) baking powder
5 ml (1 level tsp) ground mixed spice
2.5 ml ($\frac{1}{2}$ level tsp) ground cinnamon

100 g (4 oz) softened butter or soft tub margarine
175 g (6 oz) light soft brown sugar
2 eggs
75 ml (5 tbsp) milk
icing sugar, to dredge

1 GREASE A 1.6 LITRE ($2\frac{3}{4}$ pint) ring mould and scatter a third of the apple in the base.

2 PUT THE FLOUR, baking powder, spices, butter or margarine, sugar, eggs and milk in a bowl and beat until smooth.

3 FOLD IN THE remaining apple, then spoon the cake mixture into the ring mould and level the surface.

4 COOK ON HIGH for 8–9 minutes or until the cake is well risen, firm to the touch and no longer looks wet around the centre edge. Leave to cool in the dish then turn out and dredge with icing sugar. Spicy Apple Cake will keep for 1–2 days in an airtight container.

ORANGE YOGURT CAKE
——————MAKES 16 slices——————
▲

Vary this simple cake by using lemon or lime rind and decorate with similar flavoured Glacé Icing (see page 182).

175 g (6 oz) self raising flour
100 g (4 oz) caster sugar
100 g (4 oz) softened butter or soft tub margarine

finely grated rind of 1 orange
2 eggs
150 ml ($\frac{1}{4}$ pint) natural yogurt

1 GREASE A 1.1 LITRE (2 pint) ring mould and line the base with a ring of greaseproof paper.

2 PUT ALL OF the ingredients in a bowl and beat until smooth. Alternatively, put all the ingredients in a food processor or mixer and mix until smooth.

3 SPOON THE MIXTURE into the ring mould and level the surface. Cook on HIGH for 6–7 minutes or until the cake is well risen, firm to the touch and no longer looks wet around the centre edge. Leave to cool in the dish, then carefully turn out and serve cut into slices.

CHOCOLATE SWIRL CAKE
—————————SERVES 8—————————
▲

This is the quickest cake in the world to make! It is made by the all-in-one method, it is mixed in one bowl and cooks in very little time. Not only that, it looks and tastes delicious.

100 g (4 oz) soft tub margarine
100 g (4 oz) caster sugar
2 eggs

60 ml (4 tbsp) milk
175 g (6 oz) self raising flour
10 ml (2 level tsp) cocoa powder

1 GREASE A 1.6 LITRE ($2\frac{3}{4}$ pint) ring mould and line the base with a ring of greaseproof paper.

2 PUT ALL THE ingredients except the cocoa powder in a bowl and beat together until pale and fluffy.

3 SPOON HALF OF the mixture, leaving gaps between each spoonful, into the base of the ring mould.

4 BEAT THE COCOA into the remaining mixture, then spoon into the spaces left in the ring mould.

5 DRAW A KNIFE through the cake mixture in a spiral to make a marbled effect and level the surface.

6 COVER WITH ABSORBENT kitchen paper and cook on HIGH for 4–5 minutes until well risen and firm to the touch. Leave to stand for 10 minutes, then turn out and leave to cool on a wire rack.

VICTORIA SANDWICH CAKE
————————MAKES 8–10 slices————————
▲

*This is the basic cake mixture which has endless variations. The plain
version, as given here, is very pale when cooked, but if the jam is
allowed to ooze out of the sides and the top is covered with a generous
dusting of icing sugar, the cake is hard to resist.*

175 g (6 oz) self raising flour
175 g (6 oz) softened butter or soft tub
  margarine
175 g (6 oz) caster sugar

3 eggs
45 ml (3 tbsp) milk
jam, to fill
icing sugar, to dredge

1 GREASE A DEEP 20.5 cm (8 inch) round dish and line the base with
greaseproof paper.

2 PUT THE FLOUR, butter or margarine, sugar, eggs and milk in a bowl and
beat until smooth. Alternatively, put all the ingredients in a food processor
or mixer and mix until smooth.

3 SPOON THE MIXTURE into the prepared dish. Cover, stand on a roasting
rack and cook on HIGH for 6–7 minutes or until risen, slightly shrunk away
from the sides of the dish and a skewer inserted into the centre comes out
clean.

4 UNCOVER AND LEAVE to stand for 5 minutes, then turn out and leave to
cool on a wire rack.

5 WHEN COMPLETELY COLD, cut in half horizontally, then sandwich
together with jam, and dredge generously with icing sugar.

VARIATIONS
Chocolate
Replace 45 ml (3 level tbsp) flour with 45 ml (3 level tbsp) cocoa powder.
Sandwich the cakes with Vanilla or Chocolate Butter Cream (see page 183).
Coffee
Add 10 ml (2 level tsp) instant coffee granules dissolved in a little warm
water to the creamed butter and sugar mixture with the eggs, or use 10 ml
(2 tsp) coffee essence. Sandwich the cakes with Vanilla or Coffee Butter
Cream (see page 183).
Orange or lemon
Add the finely grated rind of an orange or lemon to the mixture. Sandwich
the cakes together with Orange or Lemon Butter Cream (see page 183).

Chocolate Praline Cake (page 174)

172

CHOCOLATE PRALINE CAKE
SERVES 8
▲

When making the praline, it is important to dissolve the sugar completely, before cooking to a light golden colour, to prevent it from crystallizing. Likewise, once the sugar has dissolved, do not stir but turn the bowl, in case your cooker has a hot spot.

175 g (6 oz) blanched almonds
275 g (10 oz) caster sugar
100 g (4 oz) softened butter or soft tub
   margarine
2 eggs
45 ml (3 tbsp) clear honey

150 ml ($\frac{1}{4}$ pint) soured cream
100 g (4 oz) self raising flour
40 g ($1\frac{1}{2}$ oz) cocoa powder
50 g (2 oz) ground almonds
300 ml ($\frac{1}{2}$ pint) whipping cream

1 To MAKE THE praline, grease a baking sheet and set aside. Spread out the almonds on a large ovenproof plate and cook on HIGH for 4–5 minutes until lightly browned, stirring frequently.

2 PUT 175 G (6 OZ) of the sugar in a large heatproof bowl with 60 ml (4 tbsp) water and cook on HIGH for 3 minutes. Stir until the sugar has dissolved, then continue to cook on HIGH for 5–7 minutes until the sugar is golden brown. Turn the bowl occasionally but do not stir. Add the nuts and stir until coated, then pour them on to the baking sheet. Leave to cool.

3 MEANWHILE, MAKE THE cake. Grease a 1.6 litre ($2\frac{3}{4}$ pint) ring mould and set aside.

4 PUT THE BUTTER or margarine and remaining sugar in a bowl and beat together until pale and fluffy. Gradually beat in the eggs, honey and soured cream. Fold in the flour, cocoa and ground almonds.

5 SPOON THE CAKE mixture into the prepared dish. Stand on a roasting rack and cook on HIGH for 10 minutes until risen but still slightly moist on the surface. Leave to stand for 5 minutes, then turn out on to a wire rack to cool.

6 WHILE THE CAKE is cooling, finely crush half of the praline in a coffee grinder or food processor. Coarsely crush the remainder.

7 WHIP THE CREAM until stiff, then gradually fold in the finely crushed praline. Spread the cream on to the cake to coat it completely. Sprinkle with the coarsely crushed praline.

CHOCOLATE ROLL

———SERVES 8———

▲

This is the microwave equivalent of a Swiss roll. Make sure that you have the paper ready to turn the cake on to and that you roll it up quickly or it will crack.

2 eggs, beaten
25 g (1 oz) dark soft brown sugar
50 g (2 oz) self raising flour
30 ml (2 level tbsp) cocoa powder

15 ml (1 tbsp) milk
175 g (6 oz) strawberries, hulled
300 ml ($\frac{1}{2}$ pint) double cream

1 LINE THE BASE of a shallow 23 cm (9 inch) square dish with greaseproof paper.

2 PUT THE EGGS and sugar in a medium bowl and whisk until pale and creamy and thick enough to leave a trail on the surface when the whisk is lifted. Sift in the flour and cocoa powder, then fold in using a large metal spoon. Fold in the milk.

3 POUR THE CAKE mixture into the prepared dish and level the surface. Stand on a roasting rack, cover loosely with absorbent kitchen paper and cook on HIGH for $2\frac{1}{2}$–3 minutes until slightly shrunk away from the sides of the dish, but still looks slightly moist on the surface. Leave to stand for 5 minutes.

4 MEANWHILE, PLACE A sheet of greaseproof paper on a flat surface. Turn the cake out on to the greaseproof paper and roll up with the paper inside. Leave to cool on a wire rack.

5 SLICE THE STRAWBERRIES and set aside a few slices for decoration. Whip the cream until just stiff, fold the strawberries into half of the cream and spoon the remaining cream into a piping bag fitted with a large star nozzle. Unroll the cake and spread with the strawberry and cream filling. Re-roll and place, seam side down, on a serving plate.

6 TO DECORATE, PIPE rosettes of cream the length of the roll and arrange the reserved strawberry slices on top. Serve cut into slices.

CARROT CAKE
SERVES 6–8

This is the ideal cake for cooking in the microwave as the carrots keep it very moist. Grate the carrots finely or the texture of the cooked cake will be spoilt by lumps of raw carrot.

100 g (4 oz) softened butter or soft tub
  margarine
100 g (4 oz) dark soft brown sugar
2 eggs
grated rind and juice of 1 lemon
5 ml (1 level tsp) ground cinnamon
2.5 ml (½ level tsp) ground nutmeg
2.5 ml (½ level tsp) ground cloves
15 g (½ oz) shredded coconut

100 g (4 oz) carrots, peeled and finely grated
40 g (1½ oz) ground almonds
100 g (4 oz) self raising wholemeal flour
FOR THE TOPPING
75 g (3 oz) full fat soft cheese
50 g (2 oz) icing sugar
15 ml (1 tbsp) lemon juice
25 g (1 oz) walnuts, chopped

1 GREASE A 1.6 LITRE (2¾ pint) ring mould.

2 PUT THE BUTTER or margarine and the sugar into a bowl and beat together until pale and fluffy. Add the eggs one at a time, beating well after each addition. Beat in the lemon rind and juice, spices, coconut and carrots. Fold in the ground almonds and the flour.

3 SPOON THE MIXTURE into the prepared mould and level the surface. Cover and cook on HIGH for 10 minutes. When the cake is cooked it will shrink slightly away from the sides of the mould and be firm to the touch.

4 UNCOVER AND LEAVE to stand for 10 minutes, then turn out and leave to cool on a wire rack.

5 WHEN THE CAKE is completely cold, beat together the cheese, icing sugar and lemon juice and spread it evenly over the cake, then sprinkle with the walnuts.

MARBLED APRICOT AND SESAME SEED TEABREAD
MAKES 16 slices

▲

For a good marbled effect, use a teaspoon to spoon the mixture into the dish.

225 g (8 oz) no-soak dried apricots
100 g (4 oz) plain flour
100 g (4 oz) plain wholemeal flour
5 ml (1 level tsp) baking powder
100 g (4 oz) softened butter or soft tub
  margarine

50 g (2 oz) caster sugar
2 eggs
90 ml (6 tbsp) milk
25 g (1 oz) sesame seeds
toasted sesame seeds, to decorate

1 GREASE A 1.7 LITRE (3 pint) loaf dish and line the base with greaseproof paper.

2 PUT THE APRICOTS and 150 ml ($\frac{1}{4}$ pint) water in a bowl. Cover and cook on HIGH for 4–5 minutes until the apricots are softened, stirring occasionally. Cool slightly, then purée in a blender or food processor.

3 PUT THE FLOURS, baking powder, butter or margarine, sugar, eggs and milk in a bowl and beat until smooth. Stir in the sesame seeds.

4 PUT ALTERNATE SPOONFULS of the apricot purée and the cake mixture into the prepared dish. Level the surface and sprinkle with toasted sesame seeds.

5 STAND ON A roasting rack and cook on HIGH for 8–9 minutes until risen and firm to the touch, turning the dish two or three times if rising unevenly. Leave to cool in the dish, then turn out and serve sliced.

WALNUT, BANANA AND ORANGE TEABREAD

──────────MAKES 16 slices──────────

▲

Bananas keep this teabread moist and add a delicious flavour; make sure that you mash them well before mixing with the other ingredients.

225 g (8 oz) self raising wholemeal flour
100 g (4 oz) light soft brown sugar
100 g (4 oz) butter or margarine
100 g (4 oz) walnut halves, roughly chopped
3 ripe bananas, mashed
1 egg

finely grated rind and juice of 1 large orange
2.5 ml ($\frac{1}{2}$ level tsp) ground mixed spice
FOR THE TOPPING
25 g (1 oz) walnut halves
25 g (1 oz) dried banana chips
15 ml (1 tbsp) clear honey

1 GREASE A 1.7 LITRE (3 pint) loaf dish and line with greaseproof paper.

2 PUT THE FLOUR, sugar, butter, walnuts, bananas, egg, orange rind and juice and mixed spice in a large bowl and beat thoroughly until well mixed.

3 SPOON THE MIXTURE into the prepared dish and level the surface. Sprinkle with the walnuts and banana chips for the topping. Stand on a roasting rack and cook on MEDIUM for 14 minutes until risen and firm to the touch.

4 LEAVE TO COOL in the dish. When cold turn out and brush with the honey to glaze. Serve sliced, plain or spread with a little butter or margarine. Walnut, Banana and Orange Teabread will keep wrapped in foil for 1–2 days.

FLAPJACKS
—————MAKES 16—————
▲

Few biscuits are suitable for cooking in a microwave because they can only be cooked in small batches and also often need to be turned over. These flapjacks, however, are one of the quickest and easiest biscuits to make in the microwave.

75 g (3 oz) butter or margarine
50 g (2 oz) light soft brown sugar

30 ml (2 tbsp) golden syrup
175 g (6 oz) porridge oats

1 GREASE A SHALLOW 12.5 × 23 cm (5 × 9 inch) dish.

2 PUT THE BUTTER or margarine, sugar and syrup in a large bowl. Cook on HIGH for 2 minutes until the sugar has dissolved, stirring once. Stir well, then mix in the oats.

3 PRESS THE MIXTURE into the dish. Stand on a roasting rack and cook on HIGH for 2–3 minutes until firm to the touch.

4 LEAVE TO COOL slightly, then mark into sixteen bars. Allow to cool completely before turning out of the dish.

QUEEN CAKES
—————MAKES 18—————
▲

If you want to add some colour to these pale cakes, ice them with Glacé Icing (see page 182) and decorate appropriately.

100 g (4 oz) softened butter or soft tub
  margarine
100 g (4 oz) caster sugar
2 eggs

100 g (4 oz) self raising flour
50 g (2 oz) sultanas
30 ml (2 tbsp) milk

1 PUT THE BUTTER or margarine, sugar, eggs and flour in a large bowl and beat until smooth. Alternatively, put the ingredients in a food processor or mixer and mix until smooth. Mix in the sultanas and add the milk to make a soft dropping consistency.

2 ARRANGE SIX DOUBLE layers of paper cases in a microwave muffin tray. Fill the prepared paper cases half-full and cook on HIGH for 1 minute until risen, but still slightly moist on the surface. Transfer to a wire rack to cool. Repeat twice with the remaining mixture.

VARIATIONS
Replace the sultanas with one of the following:
50 g (2 oz) chopped dates; 50 g (2 oz) chopped glacé cherries; 50 g (2 oz) chocolate chips; 50 g (2 oz) chopped crystallized ginger.

GRANARY BREAD
————MAKES one 450 g (1 lb) loaf————
▲

This bread will not have the characteristic crisp crust of conventionally baked bread because, when bread is cooked in the microwave, moisture is drawn to the surface and prevents it from becoming crisp. However, this can easily be overcome by browning the loaf under a hot grill after cooking.

450 g (1 lb) granary flour
5 ml (1 level tsp) salt

25 g (1 oz) butter or margarine
one sachet easy blend yeast

1 GREASE A 1.7 LITRE (3 pint) loaf dish.

2 PUT THE FLOUR and salt in a large bowl and rub in the butter or margarine. Stir in the yeast.

3 PUT 300 ML ($\frac{1}{2}$ PINT) water in a jug and cook on HIGH for 1–2 minutes until tepid. Pour on to the flour and mix to form a soft dough.

4 TURN OUT ON to a lightly floured surface and knead for 10 minutes or until the dough is smooth and no longer sticky. Place in a large bowl, cover with a clean tea-towel and leave to rise in a warm place for about 1 hour until doubled in size.

5 TURN THE DOUGH out on to a lightly floured surface and knead lightly until smooth. Shape the dough to fit the loaf dish, place in the dish and cover with a clean tea-towel. Leave to prove for about 30 minutes until the dough reaches the top of the dish.

6 UNCOVER, STAND ON a roasting rack and cook on HIGH for about 6 minutes or until risen and firm to the touch. Leave to stand for 10 minutes before turning out on to a wire rack to cool. If liked, cook quickly under a hot grill to brown the top.

HERB, CHEESE AND OLIVE BREAD

MAKES one loaf (16 slices)

▲

This makes a moist, quick bread, delicious served with soups or salads.

225 g (8 oz) self raising wholemeal flour
5 ml (1 level tsp) baking powder
salt and pepper
100 g (4 oz) mature Cheddar cheese, grated
45 ml (3 tbsp) roughly chopped fresh mixed
  herbs
75 g (3 oz) black olives, quartered
1 egg

30 ml (2 tbsp) olive oil
225 ml (8 fl oz) milk
FOR THE TOPPING
25 g (1 oz) mature Cheddar cheese, grated
15 ml (1 tbsp) roughly chopped fresh mixed
  herbs
25 g (1 oz) black olives, roughly chopped

1 GREASE A 1.7 LITRE (3 pint) loaf dish and line the base with greaseproof paper.

2 PUT THE FLOUR, baking powder, salt and pepper to taste, cheese, herbs, black olives, egg, olive oil and milk in a large bowl and beat until mixed.

3 SPOON THE MIXTURE into the prepared dish and level the surface. Sprinkle on the topping ingredients. Stand the dish on a roasting rack and cook on HIGH for 3 minutes, then cook on MEDIUM for a further 14 minutes or until firm to the touch and well risen.

4 LEAVE TO COOL in the dish. Serve sliced, spread with a little butter.

IRISH SODA BREAD

MAKES one large loaf (18 slices)

▲

*Serve this delicious bread warm in chunks with homemade soup, or
slice and serve with salads or stews.*

450 g (1 lb) self raising wholemeal flour plus
  extra for sprinkling
225 g (8 oz) plain flour

salt
2.5 ml ($\frac{1}{2}$ level tsp) bicarbonate of soda
600 ml (1 pint) buttermilk

1 MIX THE FLOURS, salt to taste and bicarbonate of soda in a large bowl.

2 POUR IN THE buttermilk and mix quickly to form a soft dough. Knead lightly on a floured surface and shape into a 25 cm (10 inch) round cob loaf. Cut a large cross in the top, and dust lightly with a little flour.

3 PLACE THE LOAF on a microwave baking tray or large plate and stand on a roasting rack. Cook on HIGH for 15 minutes or until risen and firm to the touch. Serve warm, or leave to cool on a wire rack and eat cold.

Flapjacks (page 178), Walnut, Banana and Orange Teabread (page 177)

BRAN MUFFINS
————MAKES 8————
▲

The perfect quick breakfast—weigh out the dry ingredients the night before and then add the liquid in the morning.

50 g (2 oz) bran
75 g (3 oz) plain wholemeal flour
7.5 ml (1½ level tsp) baking powder

1 egg, beaten
300 ml (½ pint) milk
30 ml (2 tbsp) clear honey

1 PUT THE BRAN, flour and baking powder in a bowl and mix together. Add the egg, milk and honey and stir until well mixed.

2 DIVIDE THE MIXTURE between an eight-hole microwave muffin tray and cook on HIGH for 5–6 minutes until firm to the touch.

3 LEAVE TO STAND for 5 minutes. Split each muffin in half horizontally and serve while still warm, spread with butter or margarine.

GLACÉ ICING
—MAKES about 100 g (4 oz)—
▲

The quantities given make sufficient to cover the top of an 18 cm (7 inch) cake or up to eighteen small cakes. To cover the top of a 20.5 cm (8 inch) cake, increase the quantities to 175 g (6 oz) icing sugar and 30 ml (2 tbsp) warm water. This will give a 175 g (6 oz) quantity of icing.

100 g (4 oz) icing sugar

15–30 ml (1–2 tbsp) warm water

1 SIFT THE ICING sugar into a bowl. If liked, add a few drops of any flavouring and gradually add the warm water. The icing should be thick enough to coat the back of a spoon. If necessary, add more water or sugar to adjust the consistency. Add colouring, if liked, and use at once.

VARIATIONS
Orange
Replace the water with 15 ml (1 tbsp) strained orange juice.
Lemon
Replace the water with 15 ml (1 tbsp) strained lemon juice.
Chocolate
Dissolve 10 ml (2 level tsp) cocoa powder in a little hot water and use instead of the same amount of measured water.

Coffee
Flavour with 5 ml (1 tsp) coffee essence or dissolve 10 ml (2 level tsp) instant coffee granules in a little hot water and use instead of the same amount of measured water.

Mocha
Dissolve 5 ml (1 level tsp) cocoa powder and 10 ml (2 level tsp) instant coffee granules in a little hot water and use instead of the same amount of measured water.

Liqueur
Replace 10–15 ml (2–3 tsp) of the measured water with the same amount of any liqueur.

BUTTER CREAM
————————MAKES 250 g (9 oz)————————

▲

The quantities given make sufficient to coat the sides of an 18 cm (7 inch) cake, or give a topping and a filling. If you wish to coat both the sides and give a topping or filling, increase the amounts of butter and sugar to 100 g (4 oz) and 225 g (8 oz) respectively. This will make a 350 g (12 oz) quantity.

75 g (3 oz) butter
175 g (6 oz) icing sugar

few drops of vanilla flavouring
15–30 ml (1–2 tbsp) milk or warm water

1 CREAM THE BUTTER until soft, then gradually sift and beat in the sugar, adding a few drops of vanilla flavouring and the milk or water.

Orange or lemon
Replace the vanilla flavouring with a little finely grated orange or lemon rind. Add a little juice from the fruit, beating well to avoid curdling the mixture.

Walnut
Add 30 ml (2 level tbsp) finely chopped walnuts and mix well.

Almond
Add 30 ml (2 level tbsp) finely chopped toasted almonds and mix well.

Coffee
Replace the vanilla flavouring with 10 ml (2 level tsp) instant coffee granules blended with some of the liquid, or replace 15 ml (1 tbsp) of the liquid with the same amount of coffee essence.

Chocolate
Replace 15 ml (1 tbsp) of the liquid with 25–40 g (1–$1\frac{1}{2}$ oz) chocolate, melted, or dissolve 15 ml (1 level tbsp) cocoa powder in a little hot water and cool before adding to the mixture.

Mocha
Dissolve 5 ml (1 level tsp) cocoa powder and 10 ml (2 level tsp) instant coffee granules in a little warm water taken from the measured amount. Cool before adding to the mixture.

CHAPTER 11

PRESERVING

With a microwave cooker the lengthy, laborious boiling of dangerously hot saucepans that is normally linked with making preserves is a thing of the past. The cooking is quick, ordinary heatproof bowls can be used instead of saucepans and they will be easy to clean afterwards. Microwave cookers are particularly useful for preparing small quantities of preserves.

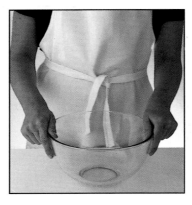

1 Always use a large bowl to avoid boiling over.

2 Use a container made of a material which withstands the high temperature of boiling sugar and always use oven gloves to remove it from the cooker.

3 Setting point of jams and marmalade is reached when a small spoonful of jam or marmalade is placed on a cold saucer and wrinkles when pushed with the tip of a finger.

4 Cook chutneys until there is no pool of liquid on the surface and the mixture is thick.

5 Sterilise all jars in the microwave. Quarter fill up to 4 jars with water, arrange in a circle in the cooker then bring to the boil on HIGH. Using oven gloves, remove each jar as it is ready and pour out the water. Invert the jar on a clean tea towel or kitchen paper and use as required.

CRUSHED STRAWBERRY JAM
MAKES 700 g (1½ lb)
▲

The advantage of cooking jam in a microwave is that you do not need to have a large quantity of fruit.

450 g (1 lb) strawberries, hulled
45 ml (3 tbsp) lemon juice

450 g (1 lb) granulated sugar
knob of butter

1 PUT THE STRAWBERRIES in a large heatproof bowl with the lemon juice. Cover and cook on HIGH for 5 minutes or until the strawberries are soft, stirring frequently.

2 LIGHTLY CRUSH THE strawberries with a potato masher. Add the sugar and stir well. Cook on LOW for 15 minutes or until the sugar has dissolved, stirring frequently.

3 COOK ON HIGH for 20–25 minutes or until a setting point is reached. Stir in the butter.

4 ALLOW THE JAM to cool slightly, then pour into hot sterilized jars. Cover and label the jam.

RASPBERRY JAM
MAKES 700 g (1½ lb)
▲

This recipe uses frozen raspberries which are available all the year round. Instructions are given for thawing them before making the jam.

450 g (1 lb) frozen raspberries
30 ml (2 tbsp) lemon juice

450 g (1 lb) granulated sugar

1 PUT THE FROZEN fruit in a large heatproof bowl and cook on HIGH for 4 minutes to thaw. Stir several times with a wooden spoon to ensure even thawing.

2 ADD THE LEMON juice and sugar. Mix well and cook on HIGH for 5 minutes until the sugar has dissolved. Stir several times during cooking.

3 COOK ON HIGH for 13 minutes until setting point is reached, stirring occasionally.

4 POUR THE JAM into hot sterilized jars. Cover and label the jam.

RHUBARB AND GINGER JAM

MAKES about 450 g (1 lb)

▲

Rhubarb is available from December to the end of June so make small batches during this time.

450 g (1 lb) rhubarb, trimmed weight
450 g (1 lb) granulated sugar
juice of 1 lemon

2.5 cm (1 inch) piece of dried root ginger, bruised
50 g (2 oz) crystallized ginger, chopped

1 CHOP THE RHUBARB into short even-sized lengths and arrange in a large heatproof bowl in layers with the sugar. Pour over the lemon juice. Cover and leave in a cool place overnight.

2 UNCOVER AND ADD the root ginger. Cook on HIGH for 5 minutes until the sugar has dissolved, stirring twice.

3 REMOVE THE ROOT ginger, add the crystallized ginger and cook on HIGH for 14 minutes or until setting point is reached.

4 POUR THE JAM into hot sterilized jars. Cover and label the jam.

DRIED APRICOT JAM

MAKES about 900 g (2 lb)

▲

It is easy to buy dried apricots so this jam can be made any time of year.

225 g (8 oz) no-soak dried apricots, roughly chopped
45 ml (3 tbsp) lemon juice

450 g (1 lb) granulated sugar
25 g (1 oz) blanched almonds, split

1 PUT THE APRICOTS, lemon juice and 600 ml (1 pint) boiling water in a large heatproof bowl. Cover and cook on HIGH for 15 minutes, stirring occasionally.

2 STIR IN THE sugar. Cook on HIGH for 2 minutes or until the sugar has dissolved. Cook on HIGH for a further 12 minutes or until setting point is reached. Stir several times during cooking. Stir in the almonds.

3 POUR THE JAM into hot sterilized jars. Cover and label the jam.

BLACKBERRY JAM
————MAKES about 700 g (1½ lb)————

▲

This is a useful jam to make when you have been blackberry picking
but only have a small quantity.

700 g (1½ lb) blackberries
45 ml (3 tbsp) lemon juice

700 g (1½ lb) granulated sugar
knob of butter

1 PUT THE BLACKBERRIES and lemon juice in a large heatproof bowl. Cover and cook on HIGH for 5 minutes until the blackberries are soft, stirring occasionally.

2 STIR IN THE sugar and cook on HIGH for 2 minutes or until the sugar has dissolved, stirring frequently.

3 COOK ON HIGH for 15 minutes or until setting point is reached. Stir in the butter.

4 POUR THE JAM into hot sterilized jars. Cover and label the jam.

GOOSEBERRY JAM
————MAKES about 900 g (2 lb)————

▲

The skins of gooseberries may be less soft than when cooked by the
conventional method but this does not spoil the finished result.

700 g (1½ lb) gooseberries
700 g (1½ lb) granulated sugar

knob of butter

1 PUT THE GOOSEBERRIES in a large heatproof bowl with 150 ml (¼ pint) water. Cover and cook on HIGH for 8–10 minutes until the gooseberries are soft, stirring frequently.

2 STIR IN THE sugar and cook on HIGH for 2 minutes until dissolved. Cook on HIGH for 20 minutes until setting point is reached. Stir in the butter.

3 POUR THE JAM into hot sterilized jars. Cover and label the jam.

ORANGE MARMALADE

MAKES about 1.1 kg (2½ lb)

▲

The skins on oranges tend to remain firm when cooked in a microwave and for this reason the rind is pared and the pith chopped.

900 g (2 lb) Seville oranges
juice of 2 lemons

900 g (2 lb) granulated sugar
knob of butter

1 PARE THE ORANGES, avoiding the white pith. Shred or chop the rind and set aside. Put the fruit pith, flesh and pips in a food processor and chop until the pips are broken.

2 PUT THE CHOPPED mixture and lemon juice in a large heatproof bowl and add 900 ml (1½ pints) boiling water. Cook on HIGH for 15 minutes.

3 STRAIN THE MIXTURE through a sieve into another large bowl and press the cooked pulp until all the juice is squeezed out. Discard the pulp. Stir the shredded rind into the hot juice and cook on HIGH for 15 minutes until the rind is tender, stirring occasionally. Stir in the sugar until dissolved.

4 COOK ON HIGH for about 10 minutes, stirring once during cooking, until setting point is reached. Stir in the butter. Remove any scum with a slotted spoon. Leave to cool for 15 minutes, then pour into hot sterilized jars. Cover and label the marmalade.

LEMON CURD

MAKES 900 g (2 lb)

▲

Making lemon curd in a microwave is much simpler than cooking it conventionally because you do not have to use a double saucepan.

finely grated rind and juice of 4 large lemons
4 eggs, beaten

225 g (8 oz) caster sugar
100 g (4 oz) butter, diced

1 PUT THE LEMON rind in a large heatproof bowl. Mix the juice with the eggs and strain into the bowl. Stir in the sugar, then add the butter.

2 COOK ON HIGH for 5–6 minutes until the curd is thick, whisking well every minute.

3 REMOVE THE BOWL from the cooker and continue whisking until the mixture is cool. Lemon curd thickens on cooling. Pour into hot sterilized jars. Cover and label. Store in the refrigerator for up to 2–3 weeks.

Griddle Scones (page 118), Crushed Strawberry Jam (page 185)

LIME CURD

-MAKES about 450 g (1 lb)-

▲

This is easy to make but it is important to whisk the curd every minute during cooking to prevent it from curdling.

finely grated rind and juice of 4 limes
3 eggs, beaten
250 g (9 oz) caster sugar

75 g (3 oz) unsalted butter, cut into small pieces

1 PUT THE LIME rind and juice in a large heatproof bowl. Gradually whisk in the eggs, sugar and butter, using a balloon whisk.

2 COOK ON HIGH for 4–6 minutes until the curd is thick, whisking well every minute.

3 REMOVE THE BOWL from the cooker and continue whisking for 3–4 minutes until the mixture is cool and thickens further.

4 POUR THE CURD into hot sterilized jars. Cover and label. The curd can be stored in the refrigerator for up to 2–3 weeks.

PICKLED CHERRIES

———MAKES about 450 g (1 lb)———

▲

Serve these sweet-sour pickled cherries as an accompaniment to cold meats or with drinks before a meal.

450 g (1 lb) cherries
225 g (8 oz) granulated sugar
300 ml ($\frac{1}{2}$ pint) white wine vinegar
4 black peppercorns

1 clove
1 bay leaf
pinch of salt

1 PUT THE CHERRIES in a 450 g (1 lb) glass jar and sprinkle with the sugar.

2 PUT THE VINEGAR and remaining ingredients in a large bowl and cook on HIGH for 6 minutes until boiling rapidly. Allow to cool, then pour the vinegar and spices over the cherries. Leave to marinate for 24 hours.

3 THE NEXT DAY, strain the vinegar from the cherries into a medium heatproof bowl. Cook the vinegar on HIGH for 5 minutes until boiling. Continue to cook on HIGH for 4 minutes until reduced slightly. Allow to cool, then return the cherries to the jars and pour over the cooled vinegar.

4 COVER THE JARS immediately with airtight and vinegar-proof tops. Store for at least 2 weeks before eating.

INDONESIAN VEGETABLE PICKLE
————————————MAKES about 700 g (1½ lb)————————————

This is an attractive mixture of crunchy vegetables, sesame seeds and nuts. Spiced pickling vinegar is available in most supermarkets or you can make your own by bringing 600 ml (1 pint) malt, wine or distilled vinegar to the boil with 50 g (2 oz) pickling spice. Leave to marinate for at least 2 hours or overnight, then strain through muslin into bottles and seal with vinegar-proof tops.

1 cm (½ inch) piece fresh root ginger, peeled and grated
2 large garlic cloves, skinned and crushed
10 ml (2 level tsp) ground turmeric
45 ml (3 tbsp) vegetable oil
150 ml (¼ pint) spiced pickling vinegar
½ cucumber
2 large carrots
175 g (6 oz) cauliflower florets
1–2 green chillies, seeded and sliced
60 ml (4 level tbsp) sesame seeds
100 g (4 oz) dark soft brown sugar
60 ml (4 tbsp) sesame seeds
100 g (4 oz) salted peanuts, roughly chopped

1 PUT THE GINGER, garlic, turmeric and oil in a large bowl and cook on HIGH for 2 minutes, stirring occasionally. Add the vinegar and cook on HIGH for 3–5 minutes or until boiling.

2 MEANWHILE, CUT THE cucumber and carrots into 0.5 cm (¼ inch) slices, and break the cauliflower into tiny florets.

3 ADD THE VEGETABLES to the boiling vinegar, cover and cook on HIGH for 2 minutes or until the liquid just returns to the boil. When boiling, cook for a further 2 minutes. Stir in the remaining ingredients and mix thoroughly together.

4 POUR INTO HOT sterilized jars. Cover with airtight and vinegar-proof tops and label. Store for 1 month before eating.

THREE PEPPER RELISH
————————MAKES about 350 g (12 oz)————————
▲

This is a very hot relish and therefore ideal served with mild flavoured foods. It will only keep for a few weeks.

1 medium onion, skinned and chopped
2 red peppers, seeded and chopped
1 red chilli, chopped
30 ml (2 tbsp) vegetable oil
2 garlic cloves, skinned and thinly sliced

15 ml (1 level tbsp) light soft brown sugar
30 ml (2 tbsp) lime juice
15 ml (1 tbsp) hoisin sauce
15 ml (1 level tbsp) paprika
pinch of salt

1 PUT ALL THE ingredients in a large bowl and mix thoroughly together.

2 COOK ON HIGH for 10–15 minutes or until the vegetables are soft, stirring occasionally.

3 POUR INTO A hot sterilized jar and cover. The relish can be stored in the refrigerator for up to 2 weeks.

SWEETCORN RELISH
————————MAKES about 700 g (1½ lb)————————
▲

For a quick version of this recipe, you can use a large can of sweetcorn kernels, drained, in place of the corn on the cob.

3 corn on the cob
2 medium onions, skinned and chopped
1 small green pepper, seeded and chopped
15 ml (1 tbsp) wholegrain mustard
5 ml (1 level tsp) ground turmeric

30 ml (2 level tbsp) plain flour
100 g (4 oz) light soft brown sugar
300 ml (½ pint) white wine vinegar
pinch of salt

1 REMOVE THE HUSK and the silk from the corn, then wrap immediately in greaseproof paper. Cook on HIGH for 8–10 minutes or until tender, turning over half way through cooking. Strip the corn from the cob.

2 PUT ALL THE remaining ingredients in a large heatproof bowl and cook on HIGH for 5–7 minutes or until boiling, stirring once.

3 ADD THE CORN to the rest of the ingredients and continue to cook on HIGH for 6–7 minutes until slightly reduced and thickened.

4 POUR INTO HOT sterilized jars. Cover with airtight and vinegar-proof tops and label.

Mango Chutney (page 195), Sweetcorn Relish

CARROT AND RAISIN CHUTNEY
──────────────MAKES about 450 g (1 lb)──────────────
▲

*This is an unusual sweet and sour chutney, good served with cold pork
or cheese.*

450 g (1 lb) carrots, coarsely grated
100 g (4 oz) raisins
15 ml (1 level tbsp) black poppy seeds
2 bay leaves
2.5 ml ($\frac{1}{2}$ level tsp) ground mixed spice

2.5 ml ($\frac{1}{2}$ level tsp) ground ginger
4 black peppercorns
50 g (2 oz) soft light brown sugar
300 ml ($\frac{1}{2}$ pint) white wine vinegar

1 PUT ALL THE ingredients in a large bowl and cook on HIGH for 12–15 minutes until the carrots are tender and the liquid has evaporated.

2 POUR INTO A hot sterilized jar. Cover with an airtight and vinegar-proof top and label.

HOT AND SPICY TOMATO CHUTNEY
──────────────MAKES about 225 g (8 oz)──────────────
▲

*This is a very quick chutney to make, not intended for storing, but to
be made and eaten within a few weeks.*

45 ml (3 tbsp) vegetable oil
3 garlic cloves, skinned and crushed
2.5 cm (1 inch) piece root ginger, peeled and
  finely grated
5 ml (1 level tsp) black mustard seeds
5 ml (1 level tsp) cumin seeds
5 ml (1 level tsp) coriander seeds

2.5 ml ($\frac{1}{2}$ level tsp) fenugreek seeds
5 ml (1 level tsp) ground turmeric
1 red chilli, seeded and finely chopped
450 g (1 lb) ripe tomatoes, skinned and
  finely chopped
salt and pepper

1 PUT THE OIL, garlic and ginger in a large bowl and cook on HIGH for 1–2 minutes, stirring once.

2 MEANWHILE, GRIND THE mustard seeds, cumin seeds, coriander seeds and fenugreek seeds in a pestle and mortar.

3 STIR ALL THE spices into the oil and cook on HIGH for 1–2 minutes, or until the spices are sizzling, stirring once.

4 ADD THE CHILLI and the tomatoes and mix thoroughly together. Cook on HIGH for 10–12 minutes, or until most of the liquid has evaporated, stirring occasionally.

5 SEASON TO TASTE with salt and pepper. Serve hot or cold. Store, covered, in the refrigerator for up to 2 weeks.

194

MANGO CHUTNEY
————MAKES about 450 g (1 lb)————
▲

*This chutney makes a delicious accompaniment to serve with roast
beef and is particularly good with a curry.*

3 mangoes
2.5 cm (1 inch) piece of fresh root ginger
1 small green chilli, seeded
100 g (4 oz) light soft brown sugar

200 ml (7 fl oz) distilled or cider vinegar
2.5 ml ($\frac{1}{2}$ level tsp) ground ginger
1 garlic clove, skinned and crushed

1 PEEL THE MANGOES and cut the flesh into small pieces. Finely chop the
ginger and chilli.

2 PUT ALL THE ingredients in a large heatproof bowl and cook on HIGH for
5 minutes or until the sugar has dissolved, stirring occasionally.

3 COOK ON HIGH for 15 minutes or until thick and well reduced. Stir two or
three times during the first 10 minutes of cooking and after every minute
for the last 5 minutes to prevent the surface of the chutney from drying out.

4 POUR INTO HOT sterilized jars, cover and label. Store for 3 months before
eating.

MIXED FRUIT CHUTNEY
————MAKES about 1.4 kg (3 lb)————
▲

*All chutneys should be stored in a cool, dry, dark place and allowed to
mature for several months before eating.*

225 g (8 oz) dried apricots
225 g (8 oz) stoned dates
350 g (12 oz) cooking apples, peeled and
    cored
1 medium onion, skinned
225 g (8 oz) bananas, peeled and sliced
225 g (8 oz) dark soft brown sugar

grated rind and juice of 1 lemon
5 ml (1 level tsp) ground mixed spice
5 ml (1 level tsp) ground ginger
5 ml (1 level tsp) curry powder
5 ml (1 level tsp) salt
450 ml ($\frac{3}{4}$ pint) distilled or cider vinegar

1 FINELY CHOP OR mince the apricots, dates, apples and onion.

2 PUT ALL THE ingredients in a large heatproof bowl and mix them
together well.

3 COOK ON HIGH for 25–30 minutes or until the mixture is thick and has
no excess liquid. Stir frequently during the cooking time, taking particular
care to stir more frequently during the last 10 minutes.

4 POUR INTO HOT sterilized jars, cover and label. Store for at least 2 months.

APPLE CHUTNEY
———MAKES about 900 g (2 lb)———

▲

This fruit chutney is good served with pork or poultry.

450 g (1 lb) cooking apples, peeled, cored
   and finely diced
450 g (1 lb) onions, skinned and finely
   chopped
100 g (4 oz) sultanas
100 g (4 oz) stoned raisins

150 g (5 oz) demerara sugar
200 ml (7 fl oz) malt vinegar
5 ml (1 level tsp) ground ginger
5 ml (1 level tsp) ground cloves
5 ml (1 level tsp) ground allspice
grated rind and juice of $\frac{1}{2}$ lemon

1 PUT ALL THE ingredients in a large heatproof bowl and cook on HIGH for
5 minutes, stirring occasionally until the sugar has dissolved.

2 COOK ON HIGH for about 20 minutes or until the mixture is thick and
has no excess liquid. Stir every 5 minutes during the cooking time to
prevent the surface of the chutney from drying out.

3 POUR INTO HOT sterilized jars, cover and label. Store for 3 months.

TOMATO CHUTNEY
———MAKES about 900 g (2 lb)———

▲

*This is a lightly spiced, smooth-textured chutney. If you have grown
an abundance of green tomatoes then these can be used instead of red
tomatoes. Green tomatoes do not have to be skinned.*

700 g (1$\frac{1}{2}$ lb) firm tomatoes
225 g (8 oz) cooking apples, peeled, cored
   and chopped
1 medium onion, skinned and chopped
100 g (4 oz) dark soft brown sugar
100 g (4 oz) sultanas

5 ml (1 level tsp) salt
200 ml (7 fl oz) malt vinegar
15 g ($\frac{1}{2}$ oz) ground ginger
1.25 ml ($\frac{1}{4}$ level tsp) cayenne pepper
2.5 ml ($\frac{1}{2}$ level tsp) mustard powder

1 PUT THE TOMATOES in a large heatproof bowl and just cover with boiling
water. Cook on HIGH for 4 minutes, then lift the tomatoes out one by one,
using a slotted spoon, and remove their skins.

2 PUT THE APPLE and onion in a blender or a food processor and blend to
form a thick paste. Coarsely chop the tomatoes.

3 MIX ALL THE ingredients together in a large heatproof bowl. Cook on
HIGH for 35–40 minutes or until the mixture is thick and has no excess
liquid. Stir every 5 minutes during cooking and take particular care,
stirring more frequently, during the last 5 minutes.

196

4 POUR INTO HOT sterilized jars, cover and label. Store for at least 2 months before eating.

SWEET INDIAN CHUTNEY
————————MAKES about 450 g (1 lb)————————

▲

Serve with very spicy Indian dishes or as a starter, Indian style, with poppadums, raita and a salad.

700 g (1½ lb) ripe tomatoes
25 g (1 oz) blanched almonds
100 g (4 oz) light soft brown sugar
4 garlic cloves, skinned and crushed
3 bay leaves

50 g (2 oz) sultanas
15 ml (1 level tbsp) nigella seeds
2.5 ml (½ level tsp) chilli powder
15 ml (3 fl oz) white wine vinegar
pinch of salt

1 ROUGHLY CHOP THE tomatoes and almonds and put into a large heatproof bowl. Add the remaining ingredients and cook on HIGH for 20 minutes or until slightly reduced and thickened.

2 POUR INTO HOT sterilized jars, cover and label.

DRIED HERBS

▲

Fresh herbs can be dried in a microwave cooker and the results are excellent because both the colour and flavour are retained. Rose petals can be dried in exactly the same way to make pot pourri.

fresh herbs such as parsley, basil, rosemary, coriander

1 STRIP THE LEAVES of the herbs off their stems and arrange in a single layer on a piece of absorbent kitchen paper.

2 COOK ON HIGH for 1 minute. Turn the leaves over and re-position and cook for a further 1–1½ minutes until the leaves are dry. You can tell when they are dry because they will crumble when rubbed between your fingers.

3 STORE IN AN airtight jar in a dark place.

CHAPTER 12

SWEET MAKING

The same principles of making sweets conventionally apply when making them in a microwave. However, sweets made in the microwave are quick and easy to make and have the advantage that they are unlikely to burn as the mixture is not in direct contact with a heat source.

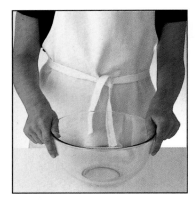

1   Use a large heatproof bowl as the container can become very hot due to its high sugar content.

2   With sugar- and syrup-based sweets, make only the quantity given in the recipe and no more, to avoid boiling over.

3   Use oven gloves when handling the bowl as the bowl can become very hot due to the conduction of heat from the mixture.

4   Stir the hot mixture with a long wooden spoon to avoid being splattered.

5   Watch chocolate carefully because if left too long in the microwave it will scorch.

6   Do not use a conventional sugar boiling thermometer but test the mixture by dropping a small ball in a glass of cold water to determine which stage the sugar has reached.

COFFEE AND WALNUT FUDGE
—————————MAKES about 20 pieces—————————
▲

It is important to make sure that the sugar has dissolved completely before boiling or the sugar will crystallize and the fudge will be grainy.

50 g (2 oz) butter
225 g (8 oz) granulated sugar
90 ml (6 tbsp) milk

45 ml (3 tbsp) coffee essence
50 g (2 oz) walnut pieces, chopped

1 OIL AN 18 × 10 CM (7 × 4 inch) container.

2 PUT THE BUTTER in a large heatproof bowl and cook on HIGH for 45 seconds or until melted. Mix in the sugar, milk and coffee essence. Cook on HIGH for 2 minutes, then stir until the sugar has dissolved.

3 COOK ON HIGH for 8 minutes without stirring, or until a teaspoonful of the mixture forms a soft ball when dropped into cold water. Turn the bowl occasionally during cooking.

4 BEAT IN THE walnuts using a wooden spoon and continue beating vigorously until the mixture is thick and creamy. (Do not continue beating after this or the fudge will become granular.)

5 POUR INTO THE prepared container. Using a sharp knife, mark into squares. Leave to set, then cut into squares when cold.

CHOCOLATE FUDGE
—————————MAKES about 36 pieces—————————
▲

This simple fudge is easy enough for children to make, just make sure that they use oven gloves when removing the bowl from the cooker.

100 g (4 oz) plain chocolate
100 g (4 oz) butter

450 g (1 lb) icing sugar
45 ml (3 tbsp) milk

1 OIL A 20.5 × 15 CM (8 × 6 inch) container.

2 PUT THE CHOCOLATE, butter, icing sugar and milk in a large heatproof bowl. Cook on HIGH for 3 minutes until the chocolate has melted.

3 BEAT VIGOROUSLY WITH a wooden spoon until the mixture is thick and creamy. (Do not continue beating after this or the fudge will become granular.)

4 POUR INTO THE container. Using a sharp knife, mark lightly into squares. Leave until set, then cut into squares when cold.

CREAMY RAISIN AND CHERRY FUDGE
MAKES about 20 pieces

▲

This fudge will keep for 2–3 weeks if stored in an airtight tin.

25 g (1 oz) butter
225 g (8 oz) granulated sugar
75 ml (5 tbsp) condensed milk

2.5 ml (½ tsp) vanilla flavouring
25 g (1 oz) seedless raisins
25 g (1 oz) glacé cherries, chopped

1 LIGHTLY OIL AN 18×10 cm (7×4 inch) container. Put the butter in a large heatproof bowl and cook on HIGH for 30 seconds, or until the butter is only just melted.

2 STIR IN THE sugar, milk and 60 ml (4 tbsp) water and continue stirring until the sugar has almost dissolved. Cook on HIGH for 2–3 minutes, then stir again to make sure all the sugar has dissolved.

3 COOK ON HIGH for 6 minutes without stirring, or until a teaspoonful of the mixture forms a soft ball when dropped into cold water. Turn the bowl occasionally during cooking.

4 CAREFULLY REMOVE THE bowl from the cooker, add the vanilla flavouring, raisins and chopped cherries and beat vigorously until the mixture is thick and creamy. (Do not continue beating after this or the fudge will become granular.)

5 Pour the fudge into the prepared container. Allow it to cool, then chill overnight before cutting it into squares.

HAZELNUT TRUFFLES
MAKES about 34

▲

These truffles work equally well using blanched almonds or walnuts instead of the hazelnuts. Vary the alcohol too; try using rum, whisky or your favourite liqueur in place of the brandy.

50 g (2 oz) hazelnuts
100 g (4 oz) plain chocolate
100 g (4 oz) unsalted butter

225 g (8 oz) icing sugar
15 ml (1 tbsp) brandy
cocoa powder, for dusting

1 SPREAD OUT THE hazelnuts on a plate and cook on HIGH for 6–8 minutes or until lightly browned, stirring frequently.

2 TIP THE NUTS on to a clean tea-towel and rub off the skins. Finely chop the nuts.

3 BREAK THE CHOCOLATE into a bowl and cook on LOW for 4 minutes or until melted, stirring occasionally. Beat in the butter, icing sugar, brandy and hazelnuts.

4 LEAVE FOR A few minutes for firm up, then shape into about 34 small balls. Dust with cocoa powder, then arrange in petit four cases. Chill in the refrigerator until firm.

RUM TRUFFLES
————MAKES about 12————
▲

Rum truffles are delicious served with coffee instead of a dessert. Remove from the refrigerator about 30 minutes before serving for maximum flavour.

50 g (2 oz) plain chocolate
25 g (1 oz) unsalted butter
50 g (2 oz) trifle sponge cakes, crumbled
25 g (1 oz) icing sugar

15 ml (1 tbsp) dark rum
25 g (1 oz) cocoa powder, icing sugar or chocolate vermicelli

1 BREAK THE CHOCOLATE into a medium bowl with the butter. Cook on LOW for 3–4 minutes, until melted, stirring occasionally. Stir in the cake crumbs, icing sugar and rum.

2 COVER AND REFRIGERATE for about 30 minutes or until the mixture is firm enough to handle.

3 LIGHTLY DUST YOUR fingers with icing sugar and roll the truffle mixture into 12 small balls, then roll each one in the cocoa powder, icing sugar or chocolate vermicelli to coat completely.

4 ARRANGE IN PETIT four cases, then chill in the refrigerator until required.

TURKISH DELIGHT
—————MAKES about 12 squares—————
▲

Store Turkish delight in an airtight container with any remaining sugar and cornflour mixture for up to 2 weeks.

25 g (1 oz) cornflour, plus extra for dusting
75 g (3 oz) granulated sugar
5 ml (1 level tsp) gelatine
few drops of rose water or peppermint
essence

few drops of red or green food colouring
(optional)
15 g ($\frac{1}{2}$ oz) icing sugar

1 LIGHTLY OIL A 13 × 10 cm (5 × 4 inch) ovenproof container. Dust generously with cornflour.

2 PUT THE SUGAR and 100 ml (4 fl oz) water in a medium heatproof bowl and cook on HIGH for 2 minutes or until hot but not boiling. Stir until the sugar has dissolved.

3 BLEND THE GELATINE with 50 ml (2 fl oz) water and set aside. Stir half of the cornflour into the sugar syrup and cook on HIGH for 1–2 minutes or until the mixture is very thick, stirring every minute. Stir in the gelatine mixture, rose water or peppermint essence to taste and a few drops of red or green food colouring, if using.

4 SPOON THE MIXTURE into the container and leave in the refrigerator for about 2 hours or until set.

5 SIFT THE REMAINING cornflour and the icing sugar on to a sheet of greaseproof paper. Turn the Turkish delight out on to it and cut into 12 squares.

6 COAT ALL THE cut surfaces of the squares with the cornflour mixture and leave, uncovered, for about 4 hours or until the surface of the Turkish delight is dry.

Creamy Raisin and Cherry Fudge (page 200)

COCONUT ICE
————MAKES 32 bars————

▲

Using a mixture of shredded and desiccated coconut gives the best texture, but if you don't have both, make up the total quantity with whichever you have.

450 g (1 lb) caster sugar
pinch of cream of tartar
150 ml ($\frac{1}{4}$ pint) milk

50 g (2 oz) shredded coconut
75 g (3 oz) desiccated coconut
few drops of red food colouring

1 PUT THE SUGAR, cream of tartar, 45 ml (3 tbsp) water and the milk in a large heatproof bowl and mix thoroughly. Cook on HIGH for 2 minutes, then stir until the sugar has dissolved.

2 COOK ON HIGH for 4–4$\frac{1}{2}$ minutes or until a teaspoonful of the syrup forms a soft ball when dropped into cold water. Shake the bowl occasionally, but do not stir the mixture or it will crystallize.

3 STIR IN THE coconut and beat thoroughly until the mixture thickens.

4 QUICKLY POUR HALF of the mixture into a 20.5 × 15 cm (8 × 6 inch) container. Colour the remaining mixture pink by adding a few drops of food colouring, then quickly spoon on top. Smooth the top and mark lightly into bars using a sharp knife. Leave until just set, then cut into bars.

PEANUT BRITTLE
————MAKES 275 g (10 oz)————

▲

Be careful when turning the bowl and when removing it from the cooker, because it will be very hot.

175 g (6 oz) caster sugar
75 ml (5 tbsp) liquid glucose
25 g (1 oz) butter

150 g (5 oz) salted peanuts
5 ml (1 level tsp) bicarbonate of soda
1.25 ml ($\frac{1}{4}$ tsp) vanilla flavouring

1 LIGHTLY OIL A large baking sheet and set aside.

2 PUT THE SUGAR, liquid glucose and 30 ml (2 tbsp) water in a large heatproof bowl. Cook on HIGH for 2 minutes until the sugar has dissolved, stirring frequently.

3 STIR IN THE butter and cook on HIGH for 1 minute or until the butter melts.

4 ADD THE PEANUTS and cook on HIGH for 6 minutes or until golden brown. Do not stir, but turn the bowl occasionally during cooking.

5 MEANWHILE, MIX TOGETHER the bicarbonate of soda, vanilla flavouring and 5 ml (1 tsp) cold water. As soon as the peanut mixture is ready, pour in the soda mixture and beat thoroughly with a wooden spoon for 2 minutes or until the mixture 'honeycombs' throughout.

6 POUR THE PEANUT mixture on to the oiled baking sheet and spread out to 0.5 cm ($\frac{1}{4}$ inch) thickness. Leave to cool.

7 WHEN THE BRITTLE is cold break into pieces. Store in an airtight tin.

CHOCOLATE CHERRY CUPS
—MAKES 12—
▲

Using a double layer of petit four cases makes them stronger and therefore easier to coat in the melted chocolate. These rich chocolates should be stored in the refrigerator and are best eaten within one week of making.

12 glacé cherries
30 ml (2 tbsp) kirsch or rum
225 g (8 oz) plain chocolate

1 egg yolk
15 ml (1 level tbsp) icing sugar, sifted

1 PUT THE CHERRIES and the kirsch or rum in a small bowl and leave to macerate for at least 1 hour.

2 ARRANGE 12 DOUBLE layers of petit four cases on a baking sheet.

3 BREAK HALF OF the chocolate into a bowl and cook on LOW for 4 minutes or until melted, stirring frequently.

4 SPOON A LITTLE chocolate into each paper case and, using a clean paint brush, coat the inside of each case with chocolate. Leave for about 30 minutes or until set. Cook the chocolate remaining in the bowl on LOW for 1–2 minutes or until just melted, then repeat the coating process, making sure the chocolate forms an even layer. Leave to set completely.

5 DRAIN THE CHERRIES, reserving the kirsch or rum. Carefully peel the paper from the chocolate shells and fill with the cherries.

6 PUT THE REMAINING chocolate into a bowl and cook on LOW for 4 minutes until melted, then add the egg yolk, icing sugar and reserved kirsch or rum. Beat well.

7 PIPE THE MIXTURE into the chocolate shells, then leave to set. Store in the refrigerator.

WHITE CHOCOLATE COLETTES
—MAKES 8—

▲

Because these chocolates contain cream they should be stored in the refrigerator until ready to eat.

75 g (3 oz) white chocolate
15 g (½ oz) butter
5 ml (1 tsp) brandy

30 ml (2 tbsp) double cream
crystallized violets, to decorate

1 ARRANGE EIGHT DOUBLE layers of petit four cases on a plate.

2 BREAK 50 G (2 OZ) of the chocolate into a small bowl and cook on LOW for 3 minutes or until just melted, stirring occasionally.

3 SPOON A LITTLE chocolate into each paper case and, using a clean paint brush, coat the inside of each case with chocolate. Leave for about 30 minutes or until set.

4 COOK THE CHOCOLATE remaining in the bowl on LOW for 1–2 minutes or until just melted, then repeat the coating process, making sure that the chocolate forms an even layer.

5 LEAVE FOR ABOUT 30 minutes in a cool place to set completely, then carefully peel away the paper from the chocolate cases.

6 BREAK THE REMAINING chocolate into small pieces and add to the bowl with the butter. Cook on LOW for 3 minutes or until just melted, stirring occasionally. Stir in the brandy. Leave for about 10 minutes, until cool and thickened slightly but not set. Whisk thoroughly.

7 WHISK IN THE cream. Leave for about 5–10 minutes or until thick enough to pipe.

8 SPOON INTO A piping bag fitted with a small star nozzle and pipe into the chocolate cases. Decorate each with a crystallized violet. Chill in the refrigerator for at least 1 hour before eating.

Chocolate Cherry Cups (page 205), White Chocolate Colettes

POPCORN
——————MAKES enough for 2 generous servings——————
▲

Specially designed microwave popcorn poppers are available, but they're very expensive and not necessary—a large bowl and a lid or heavy plate works just as well. Look out too, for bags containing popcorn and butter or salt flavourings manufactured for cooking in the microwave, they're great fun to watch as they puff up and fill with popped corn!

15 ml (1 tbsp) vegetable oil
75 g (3 oz) popping corn
25 g (1 oz) butter

30 ml (2 tbsp) clear honey or golden syrup
2.5 ml (½ level tsp) ground cinnamon
few drops of vanilla flavouring

1 PUT THE OIL in a very large heatproof bowl and cook on HIGH for 1–2 minutes or until hot. Stir in the popcorn, cover with a lid or heavy plate and cook on HIGH for 7 minutes or until the popping stops, shaking the bowl occasionally.

2 PUT THE BUTTER, the honey or syrup and the cinnamon in a heatproof jug or a small bowl and cook on HIGH for 2 minutes, until the butter has melted. Mix together, then stir in the vanilla flavouring.

3 POUR OVER THE popcorn and toss to coat completely. Best eaten while still warm.

CHOCOLATE CRACKLES
——————MAKES 24——————
▲

These all time favourites are easy enough for children to make themselves, and are especially quick and clean to prepare when using the microwave.

225 g (8 oz) plain chocolate, broken into
  small pieces
15 ml (1 tbsp) golden syrup

50 g (2 oz) butter or margarine
50 g (2 oz) cornflakes or rice breakfast cereal

1 PUT THE CHOCOLATE, golden syrup and butter or margarine in a heatproof bowl and cook on LOW for 6–7 minutes or until the chocolate has melted, stirring occasionally.

2 MIX TOGETHER, THEN fold in the cornflakes or rice cereal. When well mixed, spoon into 24 petit four cases and leave to set. Store in the refrigerator.

CHOCOLATE, FRUIT AND NUT SLICES

————————————MAKES 12 slices————————————

▲

A good sweet to make and store in the refrigerator for slicing off and nibbling as required.

40 g (1½ oz) hazelnuts
50 g (2 oz) plain chocolate, broken into small pieces
75 g (3 oz) unsalted butter
25 g (1 oz) rich tea biscuits

50 g (2 oz) glacé cherries, finely chopped
30 ml (2 tbsp) raisins
15 ml (1 tbsp) chopped mixed peel
10 ml (2 tsp) Tia Maria
30 ml (2 level tbsp) icing sugar

1 SPREAD OUT THE hazelnuts evenly on a large ovenproof plate and cook on HIGH for 2–3 minutes, or until the skins 'pop', stirring occasionally.

2 RUB OFF THE skins using a clean tea towel and chop the nuts finely.

3 PUT THE CHOCOLATE and butter in a bowl and cook on LOW for 3 minutes or until melted.

4 MEANWHILE, PUT THE biscuits into a polythene bag and crush finely using a rolling pin.

5 STIR THE CRUSHED biscuits, hazelnuts, cherries, raisins, mixed peel and Tia Maria into the chocolate mixture and mix well together. Cover and refrigerate for about 30 minutes or until the mixture is firm enough to handle.

6 TURN THE MIXTURE out on to a sheet of greaseproof paper and shape into a sausage about 25 cm (10 inches) long. Wrap up tightly in the paper, twisting the ends to keep the shape. Chill for at least 1 hour or until required.

7 UNWRAP, SIFT THE icing sugar on top and then roll in the icing sugar. Cut into 1.5 cm (¾ inch) slices to serve.

MICROWAVE THAWING AND COOKING CHARTS

ALL FOOD SHOULD BE COOKED IMMEDIATELY AFTER THAWING.

THAWING MEAT

Frozen meat exudes a lot of liquid during thawing and because microwaves are attracted to water, the liquid should be poured off or mopped up with absorbent kitchen paper when it collects, otherwise thawing will take longer. Start thawing a joint in its wrapper and remove it as soon as possible—usually after one-quarter of the thawing time. Place the joint on a microwave roasting rack so that it does not stand in liquid during thawing.

Remember to turn over a large piece of meat. If the joint shows signs of cooking give the meat a 'rest' period of 20 minutes. A joint is thawed when a skewer can easily pass through the thickest part of the meat. Chops and steaks should be re-positioned during thawing; test them by pressing the surface with your fingers—the meat should feel cold to the touch and give in the thickest part.

TYPE	TIME ON LOW OR DEFROST SETTING	NOTES
BEEF		
Boned roasting joints (sirloin, topside)	8–10 minutes per 450 g (1 lb)	*Turn over* regularly during thawing and rest if the meat shows signs of cooking. *Stand* for 1 hour.
Joints on bone (rib of beef)	10–12 minutes per 450 g (1 lb)	*Turn over* joint during thawing. The meat will still be icy in the centre but will complete thawing if you leave it to stand for 1 hour.
Minced beef	8–10 minutes per 450 g (1 lb)	*Stand* for 10 minutes.
Cubed steak	6–8 minutes per 450 g (1 lb)	*Stand* for 10 minutes.
Steak (sirloin, rump)	8–10 minutes per 450 g (1 lb)	*Stand* for 10 minutes.
LAMB/VEAL		
Boned rolled joint (loin, leg, shoulder)	5–6 minutes per 450 g (1 lb)	As for boned roasting joints of beef above. *Stand* for 30–45 minutes.
On the bone (leg and shoulder)	5–6 minutes per 450 g (1 lb)	As for beef joints on bone above. *Stand* for 30–45 minutes.
Minced lamb or veal	8–10 minutes per 450 g (1 lb)	*Stand* for 10 minutes.
Chops	8–10 minutes per 450 g (1 lb)	*Separate* during thawing. *Stand* for 10 minutes.
PORK		
Boned rolled joint (loin, leg)	7–8 minutes per 450 g (1 lb)	As for boned roasting joints of beef above. *Stand* for 1 hour.
On the bone (leg, hand)	7–8 minutes per 450 g (1 lb)	As for beef joints on bone above. *Stand* for 1 hour.
Tenderloin	8–10 minutes per 450 g (1 lb)	*Stand* for 10 minutes.
Chops	8–10 minutes per 450 g (1 lb)	*Separate* during thawing and arrange 'spoke' fashion. *Stand* for 10 minutes.
OFFAL		
Liver	8–10 minutes per 450 g (1 lb)	*Separate* during thawing. *Stand* for 5 minutes.
Kidney	6–9 minutes per 450 g (1 lb)	*Separate* during thawing. *Stand* for 5 minutes.

COOKING MEAT

TYPE	TIME/SETTING	MICROWAVE COOKING TECHNIQUE(S)
BEEF		
Boned roasting joint (sirloin, topside)	per 450 g (1 lb) Rare: 5–6 minutes on HIGH Medium: 7–8 minutes on HIGH Well done: 8–10 minutes on HIGH	*Turn* over joint halfway through cooking time. *Stand* for 15–20 minutes, tented in foil.
On the bone roasting joint (fore rib, back rib)	per 450 g (1 lb) Rare: 5 minutes on HIGH Medium: 6 minutes on HIGH Well done: 8 minutes on HIGH	*Turn* over joint halfway through cooking time. *Stand* as for boned joint.
LAMB/VEAL		
Boned rolled joint (loin, leg, shoulder)	per 450 g (1 lb) Medium: 7–8 minutes on HIGH Well done: 8–10 minutes on HIGH	*Turn* over joint halfway through cooking time. *Stand* as for beef.
On the bone (leg and shoulder)	per 450 g (1 lb) Medium: 6–7 minutes on HIGH Well done: 8–9 minutes on HIGH	*Position* fatty side down and turn over halfway through cooking time. *Stand* as for beef.
Chops	1 chop: $2\frac{1}{2}$–$3\frac{1}{2}$ minutes on HIGH 2 chops: $3\frac{1}{2}$–$4\frac{1}{2}$ minutes on HIGH 3 chops: $4\frac{1}{2}$–$5\frac{1}{2}$ minutes on HIGH 4 chops: $5\frac{1}{2}$–$6\frac{1}{2}$ minutes on HIGH	*Cook* in preheated browning dish. *Position* with bone ends towards centre. *Turn* over once during cooking.
BACON		
Joints	12–14 minutes on HIGH per 450 g (1 lb)	*Cook* in a pierced roasting bag. *Turn* over joint partway through cooking time. *Stand* for 10 minutes, tented in foil.
Rashers	2 rashers: 2–$2\frac{1}{2}$ minutes on HIGH 4 rashers: 4–$4\frac{1}{2}$ minutes on HIGH 6 rashers: 5–6 minutes on HIGH	*Arrange* in a single layer. *Cover* with greaseproof paper to prevent splattering. *Cook* in preheated browning dish if liked. *Remove* paper immediately after cooking to prevent sticking.
PORK		
Boned rolled joint (loin, leg)	8–10 minutes on HIGH per 450 g (1 lb)	As for boned rolled lamb above.
On the bone (leg, hand)	8–9 minutes on HIGH per 450 g (1 lb)	As for lamb on the bone above.
Chops	1 chop: 4–$4\frac{1}{2}$ minutes on HIGH 2 chops: 5–$5\frac{1}{2}$ minutes on HIGH 3 chops: 6–7 minutes on HIGH 4 chops: $6\frac{1}{2}$–8 minutes on HIGH	*Cook* in preheated browning dish. *Prick* kidney, if attached. *Position* with bone ends towards centre. *Turn* over once during cooking.
OFFAL		
Liver (lamb and calves)	6–8 minutes on HIGH per 450 g (1 lb)	*Cover* with greaseproof paper to prevent splattering.
Kidneys	8 minutes on HIGH per 450 g (1 lb)	*Arrange* in a circle. *Cover* to prevent splattering. *Re-position* during cooking.

THAWING POULTRY AND GAME

Poultry or game should be thawed in its freezer wrapping which should be pierced first and the metal tag removed. During thawing, pour off liquid that collects in the bag. Finish thawing in a bowl of cold water with the bird still in its bag. Chicken portions can be thawed in their polystyrene trays.

TYPE	TIME ON LOW OR DEFROST SETTING	NOTES
Whole chicken or duckling	6–8 minutes per 450 g (1 lb)	Remove giblets. *Stand* in cold water for 30 minutes.
Whole turkey	10–12 minutes per 450 g (1 lb)	Remove giblets. *Stand* in cold water for 2–3 hours.
Chicken portions	5–7 minutes per 450 g (1 lb)	*Separate* during thawing. *Stand* for 10 minutes.
Poussin, grouse, pheasant, pigeon, quail	5–7 minutes per 450 g (1 lb)	

COOKING POULTRY

TYPE	TIME/SETTING	MICROWAVE COOKING TECHNIQUE(S)
CHICKEN		
Whole chicken	8–10 minutes on HIGH per 450 g (1 lb)	*Cook* in a roasting bag, breast side down, and turn halfway through cooking. *Stand* for 10–15 minutes.
Portions	6–8 minutes on HIGH per 450 g (1 lb)	*Position* skin side up with thinner parts towards the centre. *Re-position* halfway through cooking time. *Stand* for 5–10 minutes.
Boneless breast	2–3 minutes on HIGH	
Duck		
Whole	7–9 minutes on HIGH per 450 g (1 lb)	*Turn* over as for whole chicken. *Stand* for 10–15 minutes.
Portions	4 × 300 g (11 oz) pieces: 10 minutes on HIGH, then 30–35 minutes on MEDIUM	*Position* and *re-position* as for portions above.
TURKEY		
Whole	9–11 minutes on HIGH per 450 g (1 lb)	*Turn* over three or four times, depending on size, during cooking; start cooking breast side down. *Stand* for 10–15 minutes.

THAWING FISH AND SHELLFISH

Separate cutlets, fillets or steaks as soon as possible during thawing, and remove pieces from the cooker as soon as they are thawed. Timing will depend on the thickness of the fish.

TYPE	TIME/SETTING	NOTES
Whole round fish (mullet, trout, carp, bream, whiting)	4–6 minutes per 450 g (1 lb) on LOW or DEFROST	*Stand* for 5 minutes after each 2–3 minutes. Very large fish are thawed more successfully if left to stand for 10–15 minutes after every 2–3 minutes.
White fish fillets or cutlets (cod, coley, haddock, halibut, monkfish), whole plaice or sole	3–4 minutes per 450 g (1 lb) on LOW or DEFROST	*Stand* for 5 minutes after each 2–3 minutes.
Lobster, crab, crab claws	6–8 minutes per 450 g (1 lb) on LOW or DEFROST	*Stand* for 5 minutes after each 2–3 minutes.
Crab meat	4–6 minutes per 450 g (1 lb) block on LOW or DEFROST	*Stand* for 5 minutes after each 2–3 minutes.
Prawns, shrimps, scampi, scallops	2–3 minutes per 100 g (4 oz) 3–4 minutes per 225 g (8 oz) on LOW or DEFROST	*Arrange* in a circle on a double sheet of absorbent kitchen paper to absorb liquid. Separate during thawing with a fork and remove pieces from cooker as they thaw.

COOKING FISH AND SHELLFISH

The cooking time depends on the thickness of the fish as well as the amount being cooked and whether it is cooked whole, in fillets or cut up into smaller pieces. This chart is a guide only. Always check before the end of the calculated cooking time to prevent overcooking. Simply put the fish in a single layer in a shallow dish with 30 ml (2 tbsp) stock, wine, milk or water per 450 g (1 lb) of fish (unless otherwise stated), then cover and cook as below.

TYPE	TIME/SETTING	MICROWAVE COOKING TECHNIQUE(S)
Whole round fish (whiting, mullet, trout, carp, bream, small haddock)	4 minutes on HIGH per 450 g (1 lb)	*Slash* skin to prevent bursting. *Turn* fish over halfway through cooking time if fish weighs more than 1.4 kg (3 lb). *Re-position* fish if cooking more than two.
Whole flat fish (plaice, sole)	3 minutes on HIGH per 450 g (1 lb)	*Slash* skin. Check fish after 2 minutes.
Cutlets, steaks, thick fish fillets (cod, coley, haddock, halibut, monkfish fillet)	4 minutes on HIGH per 450 g (1 lb)	*Position* thicker parts towards the outside of the dish. *Turn* halfway through cooking if steaks are very thick.
Flat fish fillets (plaice, sole)	2–3 minutes on HIGH per 450 g (1 lb)	*Check* fish after 2 minutes.
Dense fish fillets, cutlets, steaks (tuna, swordfish, conger eel), whole monkfish tail	5–6 minutes on HIGH per 450 g (1 lb)	*Position* thicker parts towards the outside of the dish. *Turn* halfway through cooking if thick.
Skate wings	6–7 minutes on HIGH per 450 g (1 lb)	*Add* 150 ml ($\frac{1}{4}$ pint) stock or milk. Cook more than 900 g (2 lb) in batches.
Smoked fish	Cook as appropriate for type of fish, e.g. whole, fillet or cutlet. See above	
Squid	Put prepared squid, cut into rings, in a large bowl with 150 ml ($\frac{1}{4}$ pint) wine, stock or water per 450 g (1 lb) of squid. Cook, covered, on HIGH for 5–8 minutes per 450 g (1 lb)	*Time* depends on size of squid—larger, older, squid are tougher and may take longer to cook.

TYPE	TIME/SETTING	MICROWAVE COOKING TECHNIQUE(S)
Octopus	Put prepared octopus, cut into 2.5 cm (1 inch) pieces, in a large bowl with 150 ml (¼ pint) wine, stock or water per 450 g (1 lb) of octopus. Cook, covered, on HIGH until liquid is boiling, then on MEDIUM for 15–20 minutes per 450 g (1 lb)	*Tenderise* octopus before cooking by beating vigorously with a meat mallet or rolling pin. *Marinate* before cooking to help tenderise. Time depends on age and size of octopus.
Scallops (shelled)	2–4 minutes on HIGH per 450 g (1 lb)	*Do* not overcook or scallops will be tough. *Add* corals for 1–2 minutes at end of cooking time.
Scallops in their shells	Do not cook in the microwave	Cook conventionally.
Mussels	Put up to 900 g (2 lb) mussels in a large bowl with 150 ml (¼ pint) wine, stock or water. Cook, covered, on HIGH for 3–5 minutes	*Remove* mussels on the top as they cook. *Shake* the bowl occasionally during cooking. *Discard* any mussels which do not open.
Cockles	Put cockles in a large bowl with a little water. Cook, covered, on HIGH for 3–4 minutes until the shells open. Take cockles out of their shells and cook for a further 2–3 minutes or until hot	Shake the bowl occasionally during cooking.
Oysters	Do not cook in the microwave	
Raw prawns	2–5 minutes on HIGH per 450 g (1 lb), stirring frequently	Time depends on the size of the prawns. *Cook* until their colour changes to bright pink.
Live lobster	Do not cook in the microwave	Cook conventionally.
Live crab	Do not cook in the microwave	Cook conventionally.
Small clams	Cook as mussels	As mussels.
Large clams	Do not cook in the microwave.	Cook conventionally.

THAWING BAKED GOODS AND PASTRY

To absorb the moisture of thawing cakes, breads and pastry, place them on absorbent kitchen paper (remove as soon as thawed to prevent sticking). For greater crispness, place baked goods and the paper on a microwave rack to allow the air to circulate underneath.

TYPE	QUANTITY	TIME ON LOW OR DEFROST SETTING	NOTES
Loaf, whole	1 large	6–8 minutes	*Uncover* and place on absorbent kitchen paper.
Loaf, whole	1 small	4–6 minutes	*Turn* over during thawing. *Stand* for 5–15 minutes.
Loaf, sliced	1 large	6–8 minutes	*Thaw* in original wrapper but remove any
Loaf, sliced	1 small	4–6 minutes	metal tags. *Stand* for 10–15 minutes.
Slice of bread	25 g (1 oz)	10–15 seconds	*Place* on absorbent kitchen paper. *Time* carefully. *Stand* for 1–2 minutes.
Bread rolls, tea-cakes, scones, crumpets etc.	2 4	15–20 seconds 25–35 seconds	*Place* on absorbent kitchen paper. *Time* carefully. *Stand* for 2–3 minutes.

Type	Quantity	Time on Low or Defrost Setting	Notes
Cakes and Pastries			
Cakes	2 small	30–60 seconds	*Place* on absorbent kitchen paper.
	4 small	1–1½ minutes	*Stand* for 5 minutes.
Sponge cake	450 g (1 lb)	1–1½ minutes	*Place* on absorbent kitchen paper. *Test* and turn after 1 minute. *Stand* for 5 minutes.
Jam doughnuts	2	45–60 seconds	*Place* on absorbent kitchen paper.
	4	45–90 seconds	*Stand* for 5 minutes.
Cream doughnuts	2	45–60 seconds	*Place* on absorbent kitchen paper.
	4	1¼–1¾ minutes	*Check* after half the thawing time. *Stand* for 10 minutes.
Cream éclairs	2	45 seconds	*Stand* for 5–10 minutes
	4	1–1½ minutes	*Stand* for 15–20 minutes.
Choux buns	4 small	1–1½ minutes	*Stand* for 20–30 minutes.
Pastry			
Shortcrust and puff	227 g (8 oz) packet	1 minute	*Stand* for 20 minutes.
	397 g (14 oz) packet	2 minutes	*Stand* for 20–30 minutes.

COOKING FROZEN VEGETABLES

Frozen vegetables may be cooked straight from the freezer. Many may be cooked in their original plastic packaging, as long as it is first slit and then placed on a plate. Alternatively, transfer to a bowl.

Vegetable	Quantity	Time on High Setting	Microwave Cooking Technique(s)
Asparagus	275 g (10 oz)	7–9 minutes	*Separate* and re-arrange after 3 minutes.
Beans, broad	225 g (8 oz)	7–8 minutes	*Stir* or *shake* during cooking period.
Beans, green cut	225 g (8 oz)	6–8 minutes	*Stir* or *shake* during cooking period.
Broccoli	275 g (10 oz)	7–9 minutes	*Re-arrange* spears after 3 minutes.
Brussels sprouts	225 g (8 oz)	6–8 minutes	*Stir* or *shake* during cooking period.
Cauliflower florets	275 g (10 oz)	7–9 minutes	*Stir* or *shake* during cooking period.
Carrots	225 g (8 oz)	6–7 minutes	*Stir* or *shake* during cooking period.
Corn-on-the-cob	1	3–4 minutes	*Do not* add water. Dot with butter, wrap in greaseproof paper.
	2	6–7 minutes	
Mixed vegetables	225 g (8 oz)	5–6 minutes	*Stir* or *shake* during cooking period.
Peas	225 g (8 oz)	5–6 minutes	*Stir* or *shake* during cooking period.
Peas and carrots	225 g (8 oz)	7–8 minutes	*Stir* or *shake* during cooking period.
Spinach, leaf or chopped	275 g (10 oz)	7–9 minutes	*Do not* add water. *Stir* or *shake* during cooking period.
Swede and Turnip, diced	225 g (8 oz)	6–7 minutes	*Stir* or *shake* during cooking period. *Mash* with butter after standing time.
Sweetcorn	225 g (8 oz)	4–6 minutes	*Stir* or *shake* during cooking period.

COOKING FRESH VEGETABLES

When using this chart add 60 ml (4 tbsp) water unless otherwise stated. The vegetables can be cooked in boil-in-the-bags, plastic containers and polythene bags—pierce the bag before cooking to make sure there is a space for steam to escape.

Prepare vegetables in the normal way. It is most important that food is cut to an even size and stems are of the same length. Vegetables with skins, such as aubergines, need to be pierced before cooking to prevent bursting. Season vegetables with salt after cooking if required. Salt distorts the microwave patterns and dries the vegetables.

VEGETABLE	QUANTITY	TIME ON HIGH SETTING	MICROWAVE COOKING TECHNIQUE(S)
Artichoke, globe	1	5–6 minutes	*Place* upright in covered dish.
	2	7–8 minutes	
	3	11–12 minutes	
	4	12–13 minutes	
Asparagus	450 g (1 lb)	7–8 minutes	*Place* stalks towards the outside of the dish. *Re-position* during cooking.
Aubergine	450 g (1 lb) 0.5 cm (¼ inch) slices	5–6 minutes	*Stir* or *shake* after 4 minutes.
Beans, broad	450 g (1 lb)	6–8 minutes	*Stir* or *shake* after 3 minutes and test after 5 minutes.
Beans, green	450 g (1 lb) sliced into 2.5 cm (1 inch) lengths	10–13 minutes	*Stir* or *shake* during the cooking period. Time will vary with age.
Beetroot, whole	4 medium	14–16 minutes	*Pierce* skin with a fork. *Re-position* during cooking.
Broccoli	450 g (1 lb) small florets	7–8 minutes	*Re-position* during cooking. *Place* stalks towards the outside of the dish.
Brussels sprouts	225 g (8 oz)	4–6 minutes	*Stir* or *shake* during cooking.
	450 g (1 lb)	7–10 minutes	
Cabbage	450 g (1 lb) quartered	8 minutes	*Stir* or *shake* during cooking.
	450 g (1 lb) shredded	8–10 minutes	
Carrots	450 g (1 lb) small whole	8–10 minutes	*Stir* or *shake* during cooking.
	450 g (1 lb) 0.5 cm (¼ inch) slices	9–12 minutes	
Cauliflower	whole 450 g (1 lb)	9–12 minutes	*Stir* or *shake* during cooking.
	225 g (8 oz) florets	5–6 minutes	
	450 g (1 lb) florets	7–8 minutes	
Celery	450 g (1 lb) sliced into 2.5 cm (1 inch) lengths	8–10 minutes	*Stir* or *shake* during cooking.
Corn-on-the-cob	2 cobs 450 g (1 lb)	6–7 minutes	*Wrap* individually in greased greaseproof paper. *Do not* add water. *Turn* over after 3 minutes.
Courgettes	450 g (1 lb) 2.5 cm (1 inch) slices	5–7 minutes	*Do not* add more than 30 ml (2 tbsp) water. *Stir* or *shake* gently twice during cooking. *Stand* for 2 minutes before draining.
Fennel	450 g (1 lb) 0.5 cm (¼ inch) slices	7–9 minutes	*Stir* and *shake* during cooking.
Leeks	450 g (1 lb) 2.5 cm (1 inch) slices	6–8 minutes	*Stir* or *shake* during cooking.
Mangetout	450 g (1 lb)	7–9 minutes	*Stir* or *shake* during cooking.

Vegetable	Quantity	Time on High Setting	Microwave Cooking Technique(s)
Mushrooms	225 g (8 oz) whole	2–3 minutes	*Do not* add water. Add 25 g (1 oz) butter or
	450 g (1 lb) whole	5 minutes	alternative fat and a squeeze of lemon juice. *Stir* or *shake* gently during cooking.
Onions	225 g (8 oz) thinly sliced	7–8 minutes	*Stir* or *shake* sliced onions. *Add only* 60 ml (4 tsbp) water to whole onions.
	450 g (1 lb) small whole	9–11 minutes	*Re-position* whole onions during cooking.
Okra	450 g (1 lb) whole	6–8 minutes	*Stir* or *shake* during cooking.
Parsnips	450 g (1 lb) halved	10–16 minutes	*Place* thinner parts towards the centre. *Add* a knob of butter and 15 ml (1 tbsp) lemon juice with 150 ml ($\frac{1}{4}$ pint) water. *Turn* dish during cooking and *re-position*.
Peas	450 g (1 lb)	9–11 minutes	*Stir* or *shake* during cooking.
Potatoes, baked jacket	1 × 175 g (6 oz) potato	4–6 minutes	*Wash* and prick the skin with a fork.
	2 × 175 g (6 oz) potatoes	6–8 minutes	*Place* on absorbent kitchen paper or napkin. *When* cooking more than two at a time arrange in a circle.
	4 × 175 g (6 oz) potatoes	12–14 minutes	*Turn* over halfway through cooking.
Potatoes, boiled (old) halved	450 g (1 lb)	7–10 minutes	*Add* 60 ml (4 tbsp) water. *Stir* or *shake* during cooking.
Potatoes, boiled (new) whole	450 g (1 lb)	6–9 minutes	*Add* 60 ml (4 tbsp) water. *Do not* overcook or new potatoes become spongy.
Sweet	450 g (1 lb)	5 minutes	*Wash* and prick the skin with a fork. *Place* on absorbent kitchen paper. *Turn* over halfway through cooking time.
Spinach	450 g (1 lb) chopped	5–6 minutes	*Do not* add water. Best cooked in roasting bag, sealed with non-metal fastening. *Stir* or *shake* during cooking.
Swede	450 g (1 lb) 2 cm ($\frac{3}{4}$ inch) dice	11–13 minutes	*Stir* or *shake* during cooking.
Turnip	450 g (1 lb) 2 cm ($\frac{3}{4}$ inch) dice	9–11 minutes	*Add* 60 ml (4 tbsp) water and *stir* or *shake* during cooking.

COOKING PASTA AND RICE

Put the pasta or rice and salt to taste in a large bowl. Pour over enough boiling water to cover the pasta or rice by 2.5 cm (1 inch). Stir and cover then cook on HIGH for the stated time, stirring occasionally.

NOTE: Large quantities of pasta and rice are better cooked conventionally.

TYPE AND QUANTITY	TIME ON HIGH SETTING	MICROWAVE COOKING TECHNIQUE
Fresh white/wholemeal/spinach pasta 225 g (8 oz)	3–4 minutes	Stand for 5 minutes. Do not drain.
Dried white/wholemeal/spinach pasta shapes 225 g (8 oz)	8–10 minutes	Stand for 5 minutes. Do not drain.
Dried white/wholemeal/spinach pasta shapes 450 g (1 lb)	12–14 minutes	Stand for 5 minutes. Do not drain.
Dried white/wholemeal spaghetti 225 g (8 oz)	7–8 minutes	Stand for 5 minutes. Do not drain.
Dried white/wholemeal spaghetti 450 g (1 lb)	8–10 minutes	Stand for 5 minutes. Do not drain.
Brown rice 225 g (8 oz)	30–35 minutes	
White rice 225 g (8 oz)	10–12 minutes	

COOKING PULSES

The following pulses will cook successfully in the microwave cooker, making considerable time savings on conventional cooking.

However, pulses with very tough skins, such as red-kidney beans, black beans, butter beans, cannellini beans, haricot beans and soya beans will not cook in less time and are better if cooked conventionally. Large quantities of all pulses are best cooked conventionally.

All pulses double in weight when cooked, so if a recipe states 225 g (8 oz) cooked beans, you will need to start with 100 g (4 oz) dried weight.

Soak beans overnight, then drain and cover with enough boiling water to come about 2.5 cm (1 inch) above the level of the beans. Cover and cook on HIGH for the time stated below, stirring occasionally.

TYPE 225 g (8 oz) quantity	TIME ON HIGH SETTING	MICROWAVE COOKING TECHNIQUE
Aduki beans	30–35 minutes	Stand for 5 minutes. Do not drain.
Black-eye beans	25–30 minutes	Stand for 5 minutes. Do not drain.
Chick peas	50–55 minutes.	Stand for 5 minutes. Do not drain.
Flageolet beans	40–45 minutes	Stand for 5 minutes. Do not drain.
Mung beans	30–35 minutes	Stand for 5 minutes. Do not drain.
Split peas/lentils (do not need overnight soaking)	25–30 minutes	Stand for 5 minutes. Do not drain.

GLOSSARY

ARCING   This happens when a dish or utensil made of metal, or with any form of metal trim, or gold or silver decoration, is used in the microwave. The metal reflects the microwaves and produces a blue spark, this is known as arcing. If this happens the cooker should be switched off immediately as arcing can damage the cooker magnetron.

ARRANGING FOOD   Arranging food in a circle with the centre left empty will provide the best results when cooking in a microwave because this allows the microwaves to penetrate from the centre as well as the outside. Unevenly shaped food such as chops, broccoli and asparagus should be arranged with the thinner parts or more delicate areas towards the centre.

AUTOMATIC PROGRAMMING   A feature which allows more than one power setting to be programmed at once, so that a number of cooking sequences can be carried out on the one setting. For example the cooker can be programmed to start off cooking the food at a HIGH setting, then to complete it on a LOW setting; to come on at a set time and cook the food so that it is ready when you come home; or thaw food and then automatically switch to a setting for cooking.

BROWNING DISHES AND GRIDDLES   These are made of a special material which absorbs microwave energy. They are heated empty in the microwave cooker for 8–10 minutes, or according to the manufacturer's instructions, during which time they get very hot. The food is then placed on the hot surface and is immediately seared and browned. Always wear oven gloves when handling as they get very hot.

BROWNING ELEMENT OR GRILL   This device works in the same way as a conventional grill and is especially useful when the microwave is the main cooking appliance since there is no need to transfer a cooked dish to a conventional grill for browning.

CLEANING   It is important to clean the interior each time it is used as any spillage will absorb microwave energy and slow down the cooking next time the cooker is used. Cleaning is easy, just wipe with a damp cloth.

COOKING TIME   Always undercook rather than overcook dishes. Overcooked food will be dry and this cannot be rectified whereas undercooked dishes can be returned to the cooker for a few extra minutes if necessary. Foods with a high moisture content will take longer to cook or reheat than drier foods, and foods which are high in fat or sugar will cook or reheat more quickly than those which are low in these ingredients. Various other factors affect the cooking times, such as whether the ingredients are warm or cold, whether you have just used the cooker and the floor is still warm, the type of cookware used or the quantity of food.

COOKWARE   You will find that much of your standard cookware is suitable for microwave use, provided it is not metal. Do not use items which are decorated with gold or silver or earthenware products that contain metal particles. Materials like ovenglass and china work well in microwave cookers but in general the best materials are those designed specifically for microwave use and which will transmit microwave energy as efficiently as possible. Although microwaves are not absorbed by the cooking dish, it may become hot during cooking because heat is conducted from the food to the container. This happens either during long periods of cooking or with foods containing a high proportion of sugar or fat. For this reason, less durable containers, such as those made of soft plastics, paper or wicker, should only be used for brief cooking times such as warming or reheating bread rolls.

COOKWARE SHAPES   The shape of your cookware is important because of the patterns in which the microwaves move around the cooker cavity. Round containers are preferable to square ones as they have no corners in which clusters of microwaves can concentrate, overcooking the food at these points. Straight sided containers are more efficient than sloping ones, which cause food at the shallower outer edge to cook more quickly. A ring-shaped container will always give the best results as the microwave energy can enter from both sides as well as the top, giving more even cooking.

COOKWARE SIZES   The depth of the container is important. Foods cooked in shallow dishes cook more quickly than those in deep dishes. Choose cooking dishes large enough to hold the quantity of food and avoid overfilling—this not only results in spillages but also prevents even cooking. Dishes should be large enough to hold foods such as fish, chicken joints or chops in a single layer.

COVERING FOOD   Food should be covered to prevent drying out. Roasting bags, absorbent kitchen paper, a plate or a lid are all suitable for covering food in the microwave. Roasting bags should be pierced or slit to allow the build-up of steam to escape during cooking. They should be tied with non-metallic ties. It is recommended that the use of cling film should be avoided in microwave cooking, as it has been found that the di-2-ethylhexyl adipate (DEHA) used to soften cling film can migrate into the food during cooking. Foil should also not be used during cooking as it can easily cause ARCING but it is useful to wrap meat during STANDING TIME.

DENSITY   A dense food such as meat will take longer to thaw, reheat or cook than porous, light and airy foods such as bread, cakes and puddings. This is because microwaves cannot penetrate as deeply into denser, heavier foods.

GRILLING   Foods such as gratins which do not brown in the microwave may be browned under a preheated grill after cooking. Remember to cook the food in a flameproof dish and not a microwave container if you intend to do this.

MEMORY CONTROLS   With a memory control it is possible to begin cooking on HIGH and then automatically switch to LOW partway through cooking time. Most cooker memories allow you to programme two or three power settings, cooking for different times or to different temperatures. Some cookers can be programmed to keep food at a required temperature for a set length of time.

MICROWAVE THERMOMETER   This is useful for cooking meat in cookers not equipped with a temperature probe and replaces a conventional meat thermometer which, because of its mercury content, cannot be used in a microwave cooker. A conventional meat thermometer should only be used _after_ food is cooked.

PRICKING AND SLASHING   Foods with a skin or membrane, such as whole fish, tomatoes, liver, egg yolks and jacket potatoes, should be pricked or slashed to prevent them bursting during cooking.

POWER OUTPUT   This refers to the wattage of the cooker. Refer to your manufacturer's handbook to find the power output of your cooker.

QUANTITIES   The larger the amount of food being cooked, the longer it will take. As a general guideline, allow about one third to one half extra cooking time when doubling the ingredients. When cooking quantities are halved, decrease the cooking time by slightly more than half the time allowed for the full quantity of that food.

REPOSITION   Foods such as meatballs should be re-arranged during cooking as foods on the outside of the dish will cook more quickly than those in the centre. Move food from the outside of the dish towards the centre, and those from the centre to the outside of the dish. Larger foods, like baked potatoes, should be turned over at the same time.

ROASTING RACK   Specially designed for use in the microwave, a microwave roasting rack is not only useful for elevating meat and poultry above their own juices during cooking, but is ideal for baking. If a cake is placed on one, the microwaves can circulate underneath the container and will allow the cake to cook more evenly.

ROTATING   Foods that cannot be stirred because it would spoil the arrangement, and foods like large cakes which cannot be repositioned or turned over, can be evenly cooked by rotating the dish once or twice during the cooking time. This is usually necessary even when the cooker has a turntable. It is particularly important if you find that cakes rise unevenly or if your cooker has hot or cold spots where food cooks at a faster or slower rate than elsewhere in the cooker.

SEASONING   Salt, if sprinkled directly on to foods such as meat, fish and vegetables, toughens and makes them dry out. It is therefore best to add salt after cooking.

STANDING TIME   Standing time is an essential part of the cooking process in which the food is usually left to stand after it has been removed from the cooker. Although the food is no longer being cooked by microwave energy, the cooking is being completed by the conduction of the heat existing in the food to the centre (if standing time were incorporated into the microwave cooking time, the outside of the food would be overcooked while the centre remained uncooked. This is because microwave energy cooks from the outside in towards the centre). Standing time will depend on the density and size of the food. Very often (as in the case of potatoes) it will take no longer than the time taken to serve the dish. However, for large joints of meat, poultry and cakes, standing time could be up to ten minutes; this time should always be followed when specified in a recipe. Meat should be wrapped in foil during standing time to keep in the heat.

STIRRER   Most microwave cookers have a built in 'stirrer' positioned behind a splatter guard or cover in the roof of the cavity. This has the same effect as a turntable and it circulates the microwaves evenly throughout the cooker.

STIRRING AND WHISKING   Since the outer edges of food normally cook first in a microwave cooker, stir from the outside of the dish towards the centre to produce an evenly cooked result.

TEMPERATURE OF FOOD   The initial temperature of the food to be cooked will affect the cooking and reheating times of all foods. Food cooked straight from the refrigerator will therefore take longer than food at room temperature. In the recipes in this book, food is at room temperature unless otherwise stated.

TEMPERATURE PROBE/FOOD SENSOR   A temperature probe is used to cook joints of meat and poultry in the microwave. It enables you to control cooking by the internal temperature of the food, rather than by time. The probe is inserted into the thickest part of the food being cooked and the other end is plugged into a special socket in the cooker cavity. The desired temperature is then selected, according to manufacturer's instructions. When the internal temperature reaches the pre-set level, the cooker switches itself off. It is, however, important that the probe is inserted in the thickest part of the flesh and not near a bone as it will give a misleading temperature reading. For this reason conventional thermometers inserted after cooking or conventional techniques for testing to see if food is cooked, are usually more reliable than probes or food sensors.

THAWING   When thawing in a microwave it is essential that the ice is melted slowly, so that the food does not begin to cook on the outside before it is completely thawed through to the centre. To prevent this happening, food must be allowed to 'rest' between bursts of microwave energy. This is especially important with large items. An AUTO-DEFROST setting does this automatically by pulsing the energy on and off, but it can be done manually by using the LOW or DEFROST SETTING if your cooker does not have an automatic defrost control.

TURNING OVER   Single items thicker than 6 cm ($2\frac{1}{2}$ inches) will cook more evenly if they are turned over once during cooking, because the microwave signal is stronger towards the upper part of the cooker. This is particularly important when the food is not covered. When turning food over, reposition so that the outside parts are placed in the centre of the dish.

TURNTABLES   To ensure even cooking, food must be turned; a turntable does this automatically. However, it is necessary to REPOSITION the food by hand. Some cookers are also equipped with automatic stirrers which are situated in the roof of the microwave. See STIRRERS.

INDEX